SEEDLING PHYSIOLOGY AND REFORESTATION SUCCESS

W0232133

FORESTRY SCIENCES

Baas P, ed: New Perspectives in Wood Anatomy. 1982. ISBN 90-247-2526-7

Prins CFL, ed: Production, Marketing and Use of Finger-Jointed Sawnwood. 1982. ISBN 90-247-2569-0

Oldeman RAA, et al., eds: Tropical Hardwood Utilization: Practice and Prospects. 1982. ISBN 90-247-2581-X

Den Ouden P and Boom BK, eds: Manual of Cultivated Conifers: Hardy in Cold and Warm-Temperate Zone. 1982. ISBN 90-247-2148-2 paperback; ISBN 90-247-2644-1 hardbound.

Bonga JM and Durzan DJ, eds: Tissue Culture in Forestry. 1982. ISBN 90-247-2660-3

Satoo T and Magwick HAI: Forest Biomass. 1982. ISBN 90-247-2710-3

Van Nao T, ed: Forest Fire Prevention and Control. 1982. ISBN 90-247-3050-3

Douglas J, ed: A Re-appraisal of Forestry Development in Developing Countries. 1983. ISBN 90-247-2830-4

Gordon JC and Wheeler CT, eds: Biological Nitrogen Fixation in Forest Ecosystems: Foundations and Applications. 1983. ISBN 90-247-2849-5

Hummel FC, ed: Forest Policy: A Contribution to Resource Development. 1984. ISBN 90-247-2883-5

Duryea ML and Landis TD, eds: Forest Nursery Manual: Production of Bareroot Seedlings. 1984. ISBN 90-247-2913-0

Manion PD, ed: Scleroderris Canker of Conifers. 1984. ISBN 90-247-2912-2

Staaf KAG and Wiksten NA, authors: Tree Harvesting Techniques. 1984. ISBN 90-247-2994-7

Duryea ML and Brown GN, eds: Seedling Physiology and Reforestation Success. 1984. ISBN 90-247-2949-1

Seedling physiology and reforestation success

Proceedings of the Physiology Working Group Technical Session

Society of American Foresters National Convention, Portland, Oregon, USA, October 16–20, 1983

edited by

MARY L. DURYEA

Oregon State University
Corvallis, USA

and

GREGORY N. BROWN

University of Maine
Orono, USA

1984 **MARTINUS NIJHOFF/DR W. JUNK PUBLISHERS**
a member of the KLUWER ACADEMIC PUBLISHERS GROUP
DORDRECHT / BOSTON / LANCASTER

Distributors

for the United States and Canada: Kluwer Academic Publishers, 190 Old Derby Street, Hingham, MA 02043, USA
for the UK and Ireland: Kluwer Academic Publishers, MTP Press Limited, Falcon House, Queen Square, Lancaster LA1 1RN, England
for all other countries: Kluwer Academic Publishers Group, Distribution Center, P.O. Box 322, 3300 AH Dordrecht, The Netherlands

Library of Congress Cataloging in Publication Data

Main entry under title:

Seedling physiology and reforestation success.

   (Forestry sciences)
   Includes bibliographical references and index.
   1. Reforestation--Congresses. 2. Trees--Seedlings--
Physiology--Congresses. I. Duryea, Mary L.
II. Brown, Gregory N. III. Society of American
Foresters. Physiology Working Group. Technical
Section. IV. Society of American Foresters. Convention
(1983 : Portland, Or.) V. Series
SD409.S413 1984   634.9'56   84--14735
ISBN-13: 978-94-009-6139-5

ISBN 978-94-009-6139-5   ISBN 978-94-009-6137-1 (eBook)
DOI 10.1007/978-94-009-6137-1

Copyright

Acknowledgments

Producing a technical session and proceedings such as this requires the coordinated efforts, diverse skills, and expertise of many people.

We appreciate the creative input of the program committee: Don Dickmann, Anne Fege, John Gordon, Donal Hook, Tom Kimmerer, Denis Lavender, Kim Steiner, and Tim White.

The technical session went very smoothly due to the efforts of George Bengtson, Don Boelter, and Steve Omi.

We are grateful for the production assistance of: Char Singkofer and Julie Cone for typing; Martha Burdick for proofing and editing; Beth Marshall for proofing and indexing; Carol Perry for editorial consultation; Gretchen Bloom for drafting and layout; and Don Poole for cover design.

We acknowledge with appreciation the financial support of the Nursery Technology Cooperative and the Department of Forest Science, Oregon State University.

Finally, we are indebted to the speaker-authors, whose tremendous dedication made the session and proceedings possible.

Contents

Planting Site and Stock Response

Matching Species and Stock Type to Site

Accelerating Early Growth in Plantations

INTRODUCTION: SEEDLING PHYSIOLOGY AND REFORESTATION SUCCESS

G. N. BROWN
Dean, College of Forest Resources, 202 Nutting Hall,
University of Maine, Orono, Maine   04469

Reforestation is a major component of silvicultural
management. Whether forest sites are to be reestablished with
the same forest species, or are to be converted to other species,
major consideration must be given to the site characteristics,
the genetic quality of the new stock, and the physiological
mechanisms of regeneration establishment and survival. Many
options are available for regeneration. In some instances,
natural regeneration, with or without advanced stocking, may
be adequate for maintaining the existing forest type. In many
instances, however, artificial regeneration is required. This
may include direct seeding, seedling planting, and asexual
cutting propagation. Many physiological mechanisms are involved
with seed germination, including the afterripening process to
break seed dormancy which may be assisted by stratification or
scarification techniques. Establishment of seedlings or cuttings
involves many physiological mechanisms related to root and
shoot development. All cases require physiological adaptation
to the stresses imposed by the environmental conditions of the
site.

Mistakes in reforestation are costly because they are
multiplied and compounded throughout the life of the forest
stand. Mistakes, of course, are represented by poor survival
and suboptimal early growth. Growth and survival of young
seedlings are functions of vigor, growth rate, and resistance
to environmental stresses: edaphic, physiographic, biological,
and climatic. Hence, proper cultural treatments and pre- and
post-planting care may solicit more effective responses than
treatments during any other stage in the life of the tree and

Duryea, M.L. and Brown, G.N. (eds.). Seedling physiology and reforestation succes
©1984, Martinus Nijhoff/Dr W. Junk Publishers, Dordrecht/Boston/London. ISBN 978-94-009-6139-5

will be manifested later in the more mature tree. Many interacting factors must be considered in selecting the proper genetic seed source, planting stock, and most effective silvicultural manipulations, including either seedbed or planting site preparation (dependent upon harvesting techniques and/or subsequent mechanical and chemical techniques) and environmental manipulation of the young stand (utilizing control of overhead forest structure and/or mechanical or chemical release from competing vegetation). All of these considerations involve an understanding of physiological mechanisms in order to make the proper decisions.

While the forester, and more specifically the silviculturist, is most concerned with "how to grow trees," the physiologist is most concerned with "how trees grow." To arrive at the most effective and ultimate solution to any problem, the problem must first be identified and understood. This involves understanding the cause and effect relationship, which means understanding physiological mechanisms in the case of regeneration problems. These mechanisms come into play during seed germination and subsequent growth processes, and include such processes as photosynthesis, respiration, nutrition, and stress relationships involving water, temperature and light. All of these processes must be considered.

The final result of reforestation reflects the composite of several sequential events. Klebs' concept (1,2) illustrates how environmental influences and hereditary characteristics merge in expression through physiological processes resulting in growth and developmental characteristics of the tree. Silvicultural decisions made at each stage of reforestation must reflect physiological concerns in order to maximize success toward this final product. For instance, on some sites, conditions may be favorable for seed germination but not favorable for subsequent seedling establishment, while on other sites the reverse may be true. These site characteristics should be identified and physiological characteristics of germinating seed and seedling growth should be identified and understood to make proper silvicultural decisions. Another example illustrating the importance of physiology to nursery practices occurs when lifting seedlings from the nursery; storage, which is often operationally necessary far in advance of planting time, may at the same time be injurious to carbohydrate reserves which may be converted or depleted during this extended storage. A third example is when the healthiest appearing seedlings may not show the best survival and growth when planted in the field, because physiological factors such as food accumulation and seasonal variation in the ability of the seedlings to produce new roots may be involved.

Current forest regeneration practices have further increased the need for an understanding of physiological mechanisms. Production of seedlings in containers is receiving much attention and opens doors for increased success in reforestation. The culturing of seedlings in containers provides the nursery manager with the opportunity to produce large and vigorous seedlings through manipulation of the soil environment and better control of adaptation during field planting. Also, the entire area of forest genetics and tree improvement is being directed toward selection and breeding to produce seed and seedlings with more vigorous growth characteristics and adaptability to various site factors. Physiological mechanisms involved in all the stages from seed production through germination and early growth must be understood for genetic manipulation to be effective in producing superior seed and planting stock.

Foresters, and more specifically silviculturists, sometimes wonder why physiological information is so important when it usually explains already known visible behavior of the tree. In the case of reforestation, I have explained some of the reasons why a better understanding of seedling physiology can enhance the ability of the silviculturist to better achieve successful reforestation. In this Proceedings, state-of-the-art information is presented on several aspects of seedling physiology and how they relate to reforestation success. An overview paper is presented on each topic followed by a specific research paper addressing an example of a current reforestation problem. Topics include propagation method which addresses clonal and tissue culture propagation and seed quality improvement; stock types which address concerns with bareroot and container-grown seedlings; matching species and stock types to site which addresses ecophysiological perspectives and internal water relations; and accelerating early growth in plantations which addresses control and effects of competing vegetation, nutrition management and mycorrhizal applications to enhance seedling survival and growth.

In conclusion, the ultimate success in management of our forests will be extensively controlled at the stage of reforestation. Reforestation success will be dependent upon making the correct decisions and applying the appropriate silvicultural techniques. Making the correct decisions and selecting the appropriate techniques will require a comprehensive understanding of the physiological processes which influence seed and cutting quality, nursery cultural practices, and seedling growth and survival on respective forest sites.

REFERENCES

1. Klebs, G. 1913. Über das Verhaltniss der Aussenwelt zur Entwicklung der Pflanzen. Sitzungsber, Heidelb. Akad Wiss., Abt. B 5, 1-47.
2. Klebs, G. 1914. Über das Treiben der einheimischen Baume, speziell der Buche. Sitzungsber. Heidelb. Akad. Wiss., Abh. Math.-Naturwiss. Kl. 3, reviewed in Plant World 18, 19(1915).

Stock Quality

Propagation Method

- **Tissue culture and vegetative propagation**—Papers 1 and 2
- **Seed**—Papers 3 and 4

1. CLONAL REFORESTATION: FORESTS OF THE FUTURE?

D.G. THOMPSON

Research Forester, International Paper Company, 34937 Tennessee
Road, Lebanon, Oregon 97355

ABSTRACT

Vegetative propagation methods make possible the production
of exact copies of trees selected for superior characteristics.
These methods include conventional techniques such as rooting
cuttings, grafting, and air-layering, as well as newer plant
tissue culture methods that include embryo or cotyledon culture,
shoot tip culture, and callus and cell culture. Rooted cuttings
have been used in reforestation programs in Japan, Europe, New
Zealand, and Canada, but not in the United States. Vegetative
propagation is easiest with young trees but becomes more difficult as trees age. Plant tissue culture methods may be useful
in propagating trees old enough to have demonstrated their
superior traits. In contrast to conventional propagation
methods in which only one copy is produced from each original
cutting, air-layer, or graft, tissue culture methods produce
several copies per culture. Only limited information is
presently available on the costs and performance of vegetatively
propagated material. Although the concept of clonal reforestation may frighten some foresters, its wise and careful use can
result in more productive and genetically diverse forests than
are now possible.

1.1 INTRODUCTION

Faced with the increasing demand for forest products including fuel and fiber, foresters have turned towards more intensive
management practices. These practices include tree breeding,
artificial regeneration, vegetation control, stand density control, fertilization, and improved harvesting and utilization
techniques, some of which will be discussed in this symposium.

Duryea, M.L. and Brown, G.N. (eds.). Seedling physiology and reforestation succes
©1984, Martinus Nijhoff/Dr W. Junk Publishers, Dordrecht/Boston/London. ISBN 978-94-009-6139-5

Although large-scale vegetative propagation of commercially important forest tree species has been used in other parts of the world for many years, with the exception of poplar, these techniques have not been used in the United States.

The goal of the paper is not to convince the reader of the benefits of vegetative propagation and clonal forestry, other recent extensive reviews attempt to do that (25, 26, 45, 47, 76), but rather to provide a state-of-the-art review of the methods of vegetative propagation: a brief historical review, conventional vegetative propagation methods (rooting cuttings, air-layering, and grafting), tissue culture propagation methods (embryo or cotyledon culture, shoot tip culture, and callus and cell culture), and finally the factors limiting the use of vegetative propagation in forestry.

1.2 NATURAL CLONAL POPULATIONS

Vegetative propagation is not quite as unnatural as one might expect. A natural stand involving one aspen (Populus tremuloides) clone, multiplied by root sprouts, occupied 43 hectares and consisted of more than 47,000 trees with an average age of 100 years (50). A black spruce (Picea mariana) clone in Northern Quebec, spread by branch layers over $14m^2$ and was estimated to be over 330 years old (62).

Stump sprouts originating from dormant buds or stool sprouts which arise from adventitious buds are common in many hardwoods and often reach merchantable size (104). Sprouting is less common in conifers but does occur in pitch pine (Pinus rigida), shortleaf pine (P. echinata), pond pine (P. serotina), bald cypress (Taxodium distichum), and coastal redwood (Sequoia sempervirens). Only the last three species form sprouts that grow to merchantable size.

Root sprouts occur in several species in addition to quaking aspen including sweetgum (Liquidambar styraciflua), beech (Fagus grandifolia), black locust (Robinia pseudoacacia), black tupelo (Nyssa sylvatica), and tamarack (Larix laricina). These species may also form clonal populations in nature.

Stream bank species, such as willow (<u>Salix</u> spp.) and poplar (<u>Populus</u> spp.), contain preformed root primordia in the stems which are important for survival and spread of these frequently flooded species (36). In nature, vegetative propagation provides an alternative method to sexual reproduction for a species to establish itself and compete with other species (20).

1.3 HISTORICAL ASPECTS

Methods for the vegetative propagation of plants by man date back to ancient times. Some woody horticultural plants, such as the "Cabernet Sauvignon" grape, have been propagated as a cloned horticultural variety for over 2,000 years. Rooting cuttings, grafting, and air-layering of woody plants were all familar to western horticulturalists in the 16th and 17th century. In Japan beginning in the 1400's, natural air-layers and rooted cuttings of Japanese cedar (<u>Cryptomeria</u> <u>japonica</u>) were used in reforestation efforts (103). Specific genotypes were selected for fast growth and adaptability and were propagated as culti-vars. Originally, mixtures of cultivars were planted, although large-scale use of the "best" clone often resulted in planta-tions of single clones. Interestingly, catastrophic failures have not occurred. Interest in rooted cuttings increased about 1910 as a result of a foliage disease that affected only seedlings (103).

In a classic paper published in 1939 titled "The Possibilities of the Clone in Forestry," Schreiner (88) discussed the use of vegetative propagation in reforestation efforts. Following World War II, improved methods for rooting of cuttings led to the development of a large-scale propagation scheme for reforesta-tion in Germany (53). Simultaneously, methods for the vegetative propagation of radiata pine (<u>Pinus</u> <u>radiata</u>) were developed in Australia and New Zealand (107). Similar systems for reforesta-tion have been developed for Norway spruce in Finland (63), Sweden (106), and Norway (91) as well as for black spruce in Canada (4), eucalyptus in Australia (40) and Brazil (110), and Sitka spruce (<u>Picea</u> <u>sitchensis</u>) in Scotland (32). There is ample evidence that vegetative propagation systems can be used

in reforestation programs but, with the exception of poplar
clones, there has not been much interest in these systems in
the United States.

1.4 VEGETATIVE PROPAGATION METHODS

Techniques for vegetative propagation of forest trees
include the more traditional methods of rooting cuttings, grafting, and air-layering (macropropagation) as well as the more
recently developed methods of plant tissue culture (micropropagation) such as the induction of adventitious buds on cultured
embryo and seedling tissues, shoot tips, and unorganized callus
and cell cultures.

1.4.1 Macropropagation

Macropropagation, as the name implies, involves the use of
a relatively large piece of plant tissue as compared to tissue
culture methods that may, in the extreme, use a single cell.
Macropropagation methods have been in use for several thousand
years, are well understood, and widely employed (39).

1.4.1.1 Rooted cuttings. Stem and shoot cuttings of many
coniferous and hardwood forest trees can be rooted (15, 27),
while in a few species, such as aspen, dormant buds on root segments can produce shoots that can then be rooted (87). The
ability to root cuttings varies with the species, but hardwoods
are generally easier to root than conifers. Exceptions include
Japanese cedar and coastal redwood, which are comparatively
easy to root, and black walnut (Juglans nigra) and the oaks
(Quercus spp.) which are difficult to root.

Rooting success depends on the time cuttings are collected,
age of the donor tree, position within the crown where the
cuttings are taken, treatment with "rooting hormones" (mainly
plant growth regulators known as auxins), length of time cuttings are stored, and to some extent, conditions under which
the cuttings are rooted (75). Age is perhaps the largest
limiting factor because rooting response is greatest in young
seedlings and declines with age. Our knowledge of the physiology
(75) and biochemistry (37) of root formation has helped make

rooted cuttings practical, but there is a great deal more we do not understand.

In West Germany, rooted cuttings of Norway spruce account for 30 to 40 percent of the total planting stock (54). Similarly, in Japan about 20 percent of the Japanese cedar currently planted is propagated by cuttings.

1.4.1.2 Air-layering. In air-layering, roots are formed on a cutting while still attached to the donor plant. This is usually accomplished by wounding the branch to be layered, treating it with auxin to stimulate root formation, and wrapping the wounded region in a medium that will retain moisture (sphagnum moss). The cutting is removed from the donor plant once the roots have begun to form.

Air-layering has been used successfully with Pinus taeda (38), P. oocarpa, P. caribaea (66), Cryptomeria japonica and Pinus patula (58). A modification of the air-layering method that girdles the stem below the wound has been successful in increasing the rooting response of air-layers of a number of difficult-to-root species. The stem girdle presumably promotes accumulation of current photosynthate in the cutting which increases food reserves and thus increases the rooting response (18, 97). The treatment of stem above the girdle with plant growth regulators (auxins, anti-auxins, and anti-gibberellins) to stimulate rooting prior to removal to a conventional rooting bench has also been helpful (38). Even though pretreatments have shown promise, their added expense has prevented their use in large-scale propagation schemes. Air-layering is used primarily when rooting cuttings is not successful and is not used more often because of its labor intensiveness, expense, and slowness.

1.4.1.3 Grafting. Grafting involves uniting pieces of two different plants together to produce a complete, new plant. This usually involves the union of a piece of the plant to be propagated (the scion) with another plant to provide the root system (the stock). The many ways the scion and stock are united have led to the existence of many types of grafts (e.g., side, cleft, and approach grafts) (39). In budding, the scion is reduced to a single shoot bud that is grafted to the stock.

The success of a graft depends on the matching of the cambia of the scion and stock. In addition to physical problems with grafts, there are genetically regulated graft incompatibilities in certain species. Douglas-fir (Pseudotsuga menziesii) is perhaps one of the best examples of graft incompatibilities (21). If an open-pollinated collection of Douglas-fir seedlings is used as rootstock, about 50 percent of the grafts will be rejected. Only by the identification and vegetative propagation by cuttings of highly graft-compatible Douglas-fir rootstock clones has the problem of graft-incompatibilities in Douglas-fir been overcome. More recently, crosses of known graft-compatible parents have resulted in graft-compatible seedling rootstock.

Grafting is a well understood method for vegetative propagation that avoids the problems of inducing root formation. It is used mostly as a method for the establishment of clonal archives and clonal seed orchards. It is particularly useful in these situations because early flowering and seed production are possible if mature tissue is used as the scion. Because it is labor intensive, grafting has not been used in reforestation programs.

1.4.2 Micropropagation

While macropropagation methods involve large pieces of tissue, micropropagation employs very small plant parts, tissues, or cells. Macropropagation methods, at best, produce one propagule from each rooted cutting, air-layer, or graft; micropropagation methods produce many propagules from each original explant. These techniques have attracted much attention because small amounts of space are required for rapid production of large amounts of material. Micropropagation methods include induction of adventitious buds on cultured cotyledons and entire embryos, induction or stimulation of adventitious or axillary buds on cultured shoot tips and regeneration of adventitious shoots and complete plants from unorganized callus and cell cultures (40, 49, 68). Although macro- and micropropagation methods appear quite different, they involve the same basic physiological principles (e.g., auxins induce root formation). For some unexplained reason, there is a correlation between easy-to-root species and ease to multiply in culture.

1.4.2.1 <u>Embryo or cotyledon cultures</u>. The use of embryonic or very young seedling tissues (less than two to three months old) has been the most successful and thus most widely used micropropagation system for both hardwood and conifer forest trees. Bud formation on these tissues was first observed almost 20 years ago (57) but was not envisioned as a way to propagate trees until the mid 1970's, and is now widely used for both hardwood and conifers (16, 22).

Shoot buds are first induced to form on cultured tissues by treatment with one or more cytokinins (a group of plant growth regulators noted for their ability to stimulate cell division and organization). Removal of cytokinin permits buds to elongate into shoots. Shoots are then treated with an auxin to induce root initiation after which removal of auxin permits root elongation. Unfortunately micropropagation multiplication rates can be seriously limited at any one of these several steps (bud induction, shoot elongation, root induction, transfer to soil). Large numbers of shoot buds have been obtained in several species [up to 180/seed in radiata pine (1); up to 264/seed (average of 97/seed) in Douglas-fir (108)], but these numbers are deceiving. In Norway spruce, Bornman (14) has shown that on average, only seven rooted, surviving propagules were produced from each original seed. Thus, although some dramatic micropropagation rates have been proposed based only on bud formation rates, they may be overly optimistic based on current techniques.

The major limitation of this method is that it can be applied only to embryonic or very young seedlings that have not had a chance to demonstrate their superior phenotypes. Embryo or cotyledon culture nevertheless has been suggested as a way to rapidly increase the limited amount of material from special crosses of superior parents or to multiply small amounts of genetically improved seed from seed orchards (92).

1.4.2.2 Shoot tip culture. Shoot tip or shoot apex cultures involve culture of the shoot apical meristem from terminal or lateral buds and generally include several to many needle or leaf primordia. Adventitious shoot buds or shoot buds in the axils of leaf or branch primordia are stimulated with cytokinins. These buds elongate into shoots which can be separated and rooted with auxins to produce an exact copy of the original selected individuals.

Hardwood trees up to 100 years old (34, 35) and conifers up to 50 years old (5, 100) have been induced to form adventitious or axillary shoot buds. Shoot tip cultures are currently the only micropropagation method that is successful with material from trees old enough to have demonstrated their superiority.

1.4.2.3 Callus and cell cultures. Callus tissue is the tissue that forms when a plant is injured and is a mass of cells growing in an unorganized manner. Callus may be grown on a solid nutrient medium or grown as a suspension of cells in liquid medium (cell cultures). Callus cultures of forest trees have been grown since the late 1930's (48) and the formation of shoot buds in elm (Ulmus campestris) (48) and coastal redwood (6) callus were reported as early as the 1950's (22).

Unorganized callus cultures may be induced to form shoot buds with cytokinin treatment, allowed to elongate into a shoot, and then treated with an auxin to form a root system. A varia- tion on this ability is known as somatic embryogenesis. In this process, non-reproductive cells (somatic cells) are induced to mimic the reproductive cells in seed tissues to form an embryo (both shoot and root form in one rather than two steps). This approach is perhaps the most attractive of the micropropa- gation methods because propagules are produced in large numbers in a single step. In theory, each cell is a potential somatic embryo. Despite such advantages, it is the least understood of the micropropagation methods.

Somatic embryogenesis has been demonstrated in several woody plants including sweetgum (Liquidambar styraciflua) (94),

Albizzia lebbeck (31), red horse chestnut (Aesculus carnea) (84),
sandalwood (Santalum album) (8, 59), Paulowina tormentosa (73),
coffee (Coffea canephora) (95), cacao (Theobroma cacao) (72),
citrus (Citrus sinensis) (56), date palm (Phoenix dactylifera)
(102), hazel (Corylus avellana) (74), and Illex aquifolium (44).
Somatic embryogenesis, however, remains to be demonstrated
among the conifers.

The regeneration of propagules from unorganized callus and
cell cultures of forest trees has received less attention than
other micropropagation methods. The main reasons are: (1) our
inability to regenerate complete propagules of many species,
and (2) that there is a greater chance for changes in the
genetic information any time plants are regenerated from an
unorganized tissue (callus or cell cultures), than when
regenerated from organized tissues (embryos, cotyledons, or
shoot tips). Such variability may at first be viewed as a
problem, but it actually can be beneficial in the production
of new genotypes. Plants from single cells of the potato
(Solanum tuberosum) clone Russet Burbank varied greatly in
their disease resistance, morphology, growth rate, and productivity (90). The exact reason for the variation of these
"somaclones" (somatic cell clones) is not known, but it provides
useful increases in genetic diversity for future breeding programs.

While embryo or cotyledon and shoot tip methods are useful
only for propagation, callus and cell culture techniques also
allow for: the regeneration of plants from haploid tissues;
the induction, selection, and regeneration of mutant plants;
the storage of valuable germplasm at low temperatures (cryogenic storage); and use of genetic engineering techniques.

The production of haploid individuals would be important
in tree breeding because haploids (or haploids that have been
induced to double their single set of chromosomes) would have
only one set of chromosomes and would not contain any hidden
recessive traits. Pollen and female gametophyte tissue (in
conifers) are naturally haploid tissues and regeneration
of haploid individuals from these tissues has shown only

limited success in conifers (11, 96) while haploid poplars
have been produced for use in breeding programs (2, 86).

Regeneration of plants from unorganized cells and tissues
also allows for the selection of specific genotypes among large
cell populations. A cell culture exposed to a mutagenizing
agent, followed by a selection factor (a disease toxin, herbicide, high salt concentration) selects for potentially resistant individuals. Selection of a salt tolerant citrus line
that regenerated salt tolerant plantlets has been demonstrated
(55). Use of this technique will increase as we learn how to
regenerate more forest tree species from callus and cell
cultures.

Valuable plant cells and tissues can also be preserved for
long periods of time by storing them at liquid nitrogen temperatures (-196°C) (85, 98). This method provides a way of preserv-
ing the diversity of a natural population in a small amount of
space for use in future breeding programs.

Perhaps the most important, yet farthest in the future,
application would be in the utilization of genetic engineering
techniques (51). These techniques depend on the ability to
regenerate complete plants from callus or cell cultures derived
from a single genetically modified plant cell.

In view of their several advantages, callus and cell culture techniques deserve more attention than is currently
devoted to them.

1.4.3 Limiting factors in vegetative propagation

Vegetative propagation has played and will continue to play
an important role in tree improvement programs. Clonal archives
and clonal seed orchards are established by grafting or rooted
cuttings. Clonal material has provided information on the
genetics of forest trees and is just beginning to be used to
eliminate the genetic variability that confounds physiological
and pathological studies (79).

Although vegetative propagation is playing an ever increasing role in reforestation programs, it does invoke certain
biological and economic limitations. The effects of aging on
rooting and performance of the propagule and the risks associated

with a narrowed genetic base are the main biological concerns. The main economic consideration involves production of vegetative propagules that are both as good or superior to seedlings and comparable in price.

1.4.3.1 <u>Maturation</u>. In most macro- and micropropagation methods, the age or maturational state of the donor plant plays a large role in the efficiency of propagation and the performance of propagules. With rooted cuttings, maturational changes result in reduced rooting ability, reduced root quality, increased time to rooting, reduced shoot growth and vigor, increased plagiotropism (continued growth of a branch cutting as a branch), and increased time required for a shoot to grow out of the plagiotropic habit.

Plagiotropism provides a good example of how maturation adversely affects vegetatively propagated material. Degree of plagiotropism varies among species, with the true firs, spruces, and Douglas-fir exhibiting a strong plagiotropic habit, the pines, especially the two needle pines, showing little if any plagiotropism, and the larches being intermediate (52). Cuttings from a Norway spruce less than 13 years old grew upright, while cuttings from trees 16 to 20 years old grew plagiotropically. The age of the donor plants also affected the time required for cuttings to grow out of the plagiotropic habit with cuttings from younger trees becoming upright earlier (81).

These maturational state effects create a problem for vegetative propagation because the precision with which superior trees can be selected increases with the age of the trees (19, 60). In Norway spruce, the best clones could not be identified at age four but could be distinguished by age eight (91). Roulund (83) has suggested that from seven to ten years are necessary for both progeny and clone testing in Norway spruce.

Because maturational changes begin to become serious problems by the age superior individuals can be recognized, attention has been directed to ways to delaying or reversing maturation. Three approaches, hedging, serial repropagation, and in vitro culture show promise for minimizing age effects.

Cuttings from the base of most woody plants root easier than those from the upper part. This positional effect, known as topophysis, appears to result from the fact that the lowest part of the tree which was formed first is the youngest or most juvenile part of the tree. Fortanier and Jonkers (29) explain this phenomenon as follows: "In plants there is no continuous replacement of cells such as occurs in animals. This explains the paradox that the first initiated, lowest, and chronologically eldest part of a seedling is most juvenile, while its more recently formed periphery is ontogenetically most mature. Full-grown and old seedlings long retain many characteristics of juvenility at their base or reproduce them when forming adventitious shoots at their base."

Repeated hedging to keep a seedling small and low often causes a tree to continue producing cuttings that are easier to root than if the seedling were allowed to grow without hedging (12, 65). Age results in an increase in the distance between the shoot apex and the root system which may account for maturational changes. Reduction of this distance may be the reason why hedging results in "rejuvenation" (78). Evidence suggests that maturation is not stopped by hedging but continues at a reduced rate so eventually the rooting ability declines (43). An additional benefit of hedging is increased branching and the large number of cuttings produced by the donor plant (80).

Serial repropagation is the repeated grafting of a scion from a mature tree onto seedling rootstock (serial regrafting) or the repeated rooting of cuttings taken from rooted cuttings of mature trees (serial rerooting) (24, 54, 67). Once again the close proximity of a mature shoot apex to a seedling root system may account for the reason why material from an old tree acts "rejuvenated" or acts young again.

The success of these methods varies with the species. Whether the scion material from the old tree is actually "rejuvenated" has not been positively demonstrated. Most claims of "rejuvenation" are based on appearance: the plant "looks juvenile" and has "juvenile-like foliage, upright growth."

Roberts and Moeller (77) suggested that regrafting may result in a "reinvigoration" (increased shoot growth) rather than a true reversal of maturation effects.

The repeated in vitro (in culture) growth of plant tissues may also be a way of altering maturational state. Franclet (30) has presented some evidence that repeated in vitro culture of redwood shoot tips from an old tree reduced the plagiotropic growth of the propagules. Perhaps better evidence comes from work with the grape (Vitis vinifera) cultivar "Cabernet Sauvignon." Shoot tips from a mature vine were grown in culture and produced shoots that exhibited seedling characteristics (presence of tendrils and typical seedling leaf patterns) whereas grafts or rooted cuttings retained the adult form (70).

Work with English ivy (Hedera helix) also suggests that somatic embryogenesis may be another in vitro way of "rejuvenating" mature material. English ivy exhibits distinctive juvenile (heart-shaped leaves, trailing growth habit, and frequent roots along stem) and adult (oval leaves, upright habit, and no roots along stem) forms. The adult tissues were induced to form callus that formed somatic embryos which produced plantlets with distinctly juvenile characteristics, while the callus from juvenile tissues could only be induced to form shoots which exhibited juvenile characteristics (7). Further work is necessary to discern the effectiveness of all of these "rejuvenation" techniques.

The absence of markers for juvenile and mature states other than morphological characteristics limits our ability to determine if true "rejuvenation" has occurred. Changes in leaf morphology may simply be a result of our treatments and may not indicate an actual reversal of the maturation state. Certainly, a better understanding of the physiology and biochemistry of maturation would be helpful (10, 12).

The problems of maturation can be avoided altogether either by propagating young, untested material or by early selection of superior individuals before maturation becomes a problem. The propagation of untested, juvenile material has been successfully used to multiply known superior provenances, crosses, or progenies without individual selection and is known as "bulk propagation" (83). In bulk propagation, there is no selection of individual clones so the genetic variability is kept high. Bulk propagation is presently used to propagate genetically improved black spruce in Canada (4) and Norway spruce in Sweden (106).

Methods for the identification of superior individuals at an early age are not well developed at present. Selection and testing of individuals would allow the propagation of the best individuals ("multiclonal varieties") (83) but invokes the risk of reducing genetic variability. As a propagation method, it is not as widely used as bulk propagation, mainly because of maturational problems. Multiclonal varieties have been very successful with eucalyptus in Brazil (110) where juvenile stump sprouts of selected mature trees are rooted and used to establish plantings.

1.4.3.2 <u>Growth and form</u>. Most of the studies to compare the growth and performance of vegetatively propagated material have involved comparisons between rooted cuttings and seedlings. Results vary with some finding cuttings inferior to seedlings (89, 99), some equal (3, 28), and some finding cuttings superior to seedlings (3, 83). Conclusions drawn from such studies depends to a large degree on the maturational state of the donor plant and differences in size and the quality of the stock at the time of planting (3, 99).

Results of work with cuttings of radiata pine that were either hedged (retarded maturation rate) or allowed to grow into trees (normal maturation rate) showed similar growth rates, but hedge-origin cuttings produced trees with larger volumes, more and larger lower branches (lessened self pruning), fewer cones, more stem taper, and poorer form than cuttings from tree-form clones after seven growing seasons. Rooted cuttings from mature trees resulted in smaller trees with better bole characteristics, with higher quantity wood than cuttings from hedged trees which produced more wood volume but of a lower quality (9, 43). Cuttings from mature trees of radiata pine also suffered more browse damage than younger material (28, 105).

Small plantings of trees produced by micropropagation have been established and information on their growth and performance is just becoming available. Loblolly pine propagules regenerated from embryonic tissues grow at rates similar to seedlings from the same seed families or seed lots (61). Propagules produced by shoot tip culture from 20-year-old Eucalyptus citriodora trees grew to a height of 6 m in 18 months (34). Propagules of 100-year-old teak trees also propagated by shoot tip culture are also under study (35). Additional information on the performance of micropropagated material is needed.

1.4.3.3 Cost of vegetative propagation. The economics of clonal reforestation is an area that has not received adequate attention. Many workers simply believe Kleinschmit's (53) general statement that to be economical, cuttings cannot cost more than two to three times the cost of seedlings (63, 107) and do not attempt to analyze their propagation costs. Zobel (109) has argued that comparisons between vegetative propagules and seedlings are not useful because they often ignore the added genetic gain captured by vegetative propagules. He also suggests that a few percentage points of added gain can justify the added cost. Propagation costs, nevertheless, are an important consideration.

Some cost estimates for rooted cuttings have been published. In 1968, rooted cuttings of up to 15-year-old radiata pine in New Zealand were estimated to cost N.Z.$20/1,000 as compared to N.Z.$7/1,000 for seedlings (101). In the same species in 1974/1975, rooted cuttings could be produced for N.Z.$88/1,000 (increased cost was due to girdling of stems before rooting) (33). In Canada, black spruce cuttings of seedlings could be produced for Can.$100/1,000 as compared to Can.$110/1,000 for 1-1 bareroot transplants (75). These results all demonstrate that the costs to produce rooted cuttings are comparable but are usually more expensive than seedlings. Only by increasing rooting or multiplication rates can the cost per propagule be reduced. Rauter (75) suggests that the most promising are (1) easy-to-root species such as willow

and poplar, (2) rooting of juvenile material, and (3) micropropagation methods.

Costs to produce forest trees by micropropagation are just
beginning to become available. Radiata pine in New Zealand
propagated from embryonic tissues have been valued at N.Z.$290/
1,000 as compared to N.Z.$50/1,000 for 1-0 bareroot seedlings
or almost six times the seedling costs (93). Brown and Sommer
(16) have estimated the cost to multiply sweetgum by a shoot
tip system to be U.S.$123/1,000 (not assuming production on a
commercial scale) as compared to U.S.$65/1,000 for 1-0 bareroot
nursery stock and U.S.$75/1,000 for container-grown seedlings.
Because labor costs are high in micropropagation systems, it
is believed that large-scale operations and less labor-inten-
sive methods such as somatic embryogenesis will reduce micropropagation costs.

1.4.3.4 Deployment of clones. There are risks associated
with clonal forestry that seem overwhelming because we have
such little experience managing this type of forest. The risks
of clonal forestry can be counterbalanced with the resulting
gains (23, 109). These risks primarily arise from restricting
the genetic diversity of plantations when reducing the number
of clones from many to one or a very few. Zobel (109) believes
that "Great care needs to be taken in invoking the horrors of
monoculture, but monoculture can be a horror if ignored."

Many people believe that clones have little or no adaptability, while in fact any genotype has a certain ability to
adapt to pests and adverse conditions. The genetic diversity
of a clonal plantation is directly under the control of the
forester to make as wide or narrow as desired. One should
remember that a seedling plantation consists of a continuum
of genotypes in which a pest may quickly adapt to similar
genotypes and affect more individuals. In the clonal plantation, a diverse collection of clones may be so unrelated as to
prevent adaptation of pests to other genotypes (64). One must
also remember that genetic diversity is no guarantee against
pests as illustrated by the virtual lack of resistance in

American elm (Ulmus americana) or American chestnut (Castanea dentata) to exotic pathogens.

A "safe" number of clones depends on the rotation age, intensity of management, genetic variability of the species, genetic diversity of the clones, and the acceptable risk and loss levels of that particular situation (23, 109). It is commonly assumed that the larger the number of clones, the lower the risk of loss. Libby (64) has presented theoretical evidence that suggests the probability of an unacceptable loss rate increases as the number of clones increases from one to two, increases further with three clones, but begins to decline with four clones. Libby (64) further concluded that from 7 to 30 clones would be safe for use in a "typical" clonal plantation. Actual data on this question should now be sought.

Concern over the number of clones necessary to make a clonal plantation safe had led to establishment of government regulations of clonal forestry in several European countries. In Germany, since 1977 clonal plantations must consist of mixtures of tested clones in specified proportions (multiclonal propagation) (69). Sweden has developed similar regulations that allow for fewer clones as more information on clonal performance becomes available (41).

Exactly how these clones should be handled in the forest has also been discussed. Early suggestions (88) were that clones should be planted in random mixtures. Although there is little direct evidence, it is thought that clones may compete among themselves for light, water, and nutrients at the expense of the total productivity of the stand. This had led to the idea that mosaics of small blocks of single clones should be planted (23, 43, 64, 109). In such stands, thinning, harvesting, or replacement could be done with ease in the event that certain clones failed, matured early, or did not perform to standards.

1.5 CONCLUSIONS

Although great advances have been made in the use of vegetative propagation methods for forest trees since the 1974

IUFRO meeting in New Zealand, some of the same problems face us almost ten years later. Many biological limitations have been overcome by the use of juvenile material, but we still face adverse maturational effects in multiclonal systems. The risks of monoculture remains a large public concern about clonal reforestation.

Clonal reforestation is now a reality through the use of rooted cuttings of juvenile material. Micropropagation systems still require further refinement and testing before they can be considered for operational use. Debates still continue over the advantages and disadvantages of "bulk" and "multi-clonal" propagation schemes. The majority of clonal reforestation programs involve bulk propagation. Although multiclonal propagation promises greater gains with material that has been tested, their use depends on our ability to overcome the adverse effects of maturation. Flexibility in the use of either bulk or multiclonal propagation methods is important in order to propagate the best material currently available, be it seed of special crosses, seed orchard seed, or mature individuals.

It will be difficult to replace a method that works as well as seed propagation with a less familiar method such as vegetative propagation (17). Much can be accomplished with a slow, gradually developing, well-designed program that avoids exagerated promises. The close linking of breeding programs and clonal reforestation programs is essential for continued genetic improvement. Although forest geneticists are perhaps the group most interested in clonal forestry at this time, clonal forestry should become an integration of genetics, physiology, pathology, silviculture, and ecology.

Conventional macropropagation methods such as rooted cuttings, grafting, and air-layering are well understood. Among the micropropagation methods, more attention needs to be paid to shoot tip culture, regeneration from callus and cell cultures, and perhaps most importantly somatic embryogenesis. Greater attention needs to be paid to the documentation of realistic propagation rates possible with micropropagation systems. The effects of maturation on both macro- and micropropagation

methods requires more attention. Techniques that show promise
in delaying or reversing maturation such as serial regrafting,
serial rerooting, and in vitro culture require more study before
their effectiveness can be determined. Good data on the benefits, economics, and performance of the propagules are essential. Information on the deployment of clonal material including the safe number of clones and their arrangement in either
mixtures or blocks is needed.

Clonal reforestation programs must recognize public concerns
of the risks associated with clones and monoculture and should
attempt to educate the public and government officials by demon-
stration plantations. Nanson (71) has suggested the use of
comparison plantations including selected and unselected clones
along with seedlings to confirm performance and to demonstrate
the diversity possible with vegetative propagules. Regulation
of use is inevitable in light of the European experience.
Rather than being forced to accept regulations devised by
politicians, foresters should develop their own guidelines in
advance following the European pattern.

Just as methods for breeding trees, artificial regeneration,
vegetation control, and forest fertilization were considered
impractical only 40 years ago, so large-scale vegetative propagation seems today in the United States. It is interesting to
note that northern Europe, which 20 to 30 years ago began the
first forest tree breeding and forest fertilization programs,
is now developing the option of clonal reforestation. Only
recently has Canada begun using these techniques while in the
United States no similar programs have yet been implemented with
the exception of poplar clones.

The clone does have a role in forestry as demonstrated by
programs in West Germany, Norway, Finland, Sweden, Scotland,
Brazil, Australia, New Zealand, and Canada. The question is
no longer will clonal forestry provide at least some of the .
future forests in the United States, but rather when will it
become prevalent.

REFERENCES

1. Aitken, J., K.J. Horgan and T.A. Thorpe. 1981. Influence of explant selection on the shoot-forming capacity of juvenile tissue of Pinus radiata. Can. J. For. Res. 11: 112-7.

2. Anonymous. 1975. Induction of haploid poplar plants from anther culture in vitro. Scientia Sinica, 18:769-78.

3. Anonymous. 1981. Growth of juvenile cuttings compared with seedlings. In Forest Research Institute Report 1981, New Zealand Forest Service, Wellington, New Zealand. 120 p. (p. 18).

4. Armson, K.A., M. Fung and W.R. Bunting. 1980. Operational rooting of black spruce cuttings. J. For. 78(6): 341-3.

5. von Arnold, S. and T. Eriksson. 1979. Induction of adventitious buds on Norway spruce (Picea abies) grown in vitro. Physiol. Plant. 45:29-34.

6. Ball, E. 1950. Differentiation in a callus culture of Sequoia sempervirens. Growth, 14:295-325.

7. Banks, M.S. 1979. Plant regeneration from callus from two growth phases of English ivy, Hedera helix L. Z. Pflanzenphysiol. 92:349-53.

8. Bapat, V.A. and P.S. Rao. 1979. Somatic embryogenesis and plantlet formation in tissue cultures of sandalwood. Ann. Bot. 44:629-30.

9. Bolstad, P.V. and W.J. Libby. 1982. Comparisons of radiata pine cuttings of hedge and tree form origin after seven growing seasons. Silv. Gen. 31(1):9-13.

10. Bonga, J.M. 1980. Plant propagation through tissue culture, emphasizing woody species. In: F. Sala, B. Parisi, R. Cella, and O. Ciferri, eds. Plant cell cultures: Results and perspectives. Elsevier/North Holland Biomedical Press, Amsterdam. 428 p. (p. 254-64).

11. Bonga, J.M. 1981. Haploid tissue culture and cytology of conifers. In: 46(p. 283-93).

12. Bonga, J.M. 1982. Vegetative propagation in relation to juvenility, maturity and rejuvenation. In: 13 (p. 387-412).

13. Bonga, J.M. and D.J. Durzan. 1982. Tissue culture in forestry. Martinus Nijhoff, The Hague. 420 p.

14. Bornman, C.H. 1983. Possibilities and constraints in the regeneration of trees from cotyledonary needles of Picea abies in vitro. Physiol. Plant. 57:5-16.

15. Brix, H. and R. van den Driessche. 1977. Use of rooted cuttings in reforestation. B.C. Forest Service/Canadian For. Service Joint Report #6. 16 p.

16. Brown, C.L. and H.E. Sommer. 1982. Vegetative propagation of dicotyledonous trees. In: 13(p. 109-49).

17. Burdon, R.D. 1982. The roles and optimal place of vegetative propagation in tree breeding strategies. In: 47 (p. 66-83).

18. Cameron, R.J. 1970. Translocation of carbon-14-labelled assimilates in shoots of Pinus radiata: the effects of girdling and indole butyric acid. J. Exp. Bot. 21(69) 943-50.

19. Cannell, M.G.R. and R.C.B. Johnstone. 1978. Free or lammas growth and progeny performance in *Picea sitchensis*. Silv. Gen. 27(6):248-54.
20. Cook, R.E. 1983. Clonal plant populations. Am. Scientist, 71(3):244-53.
21. Copes, D.L. 1973. Inheritance of graft compatibility in Douglas-fir. Bot. Gaz. 134(1):49-52.
22. David, A. 1982. In vitro propagation of gymnosperms. In: 13(p. 72-108).
23. Dinus, R.J. 1982. Pests and intensive culture: problems and prospects. In: B.A. Thielges, ed., Proceedings of the 7th North American Forest Biology Workshop, July 26-28, 1982, Lexington, Kentucky. (p. 82-97).
24. Dormling, I. and H. Kellerstam. 1981. Rooting and rejuvenation in propagating old Norway spruce by cuttings. In: 26 (p. 65-72).
25. Eliasson, L. 1977. Vegetative propagation of forest trees - physiology and practice. Institute for Forest Improvement and Department of Forest Genetics, College of Forestry, the Swedish University of Agricultural Sciences, 159 p.
26. Eriksson, G. and K. Lundkuist. 1981. Symposium on clonal forestry. Department of Forest Genetics, the Swedish University of Agricultural Sciences, 131 p.
27. Farmer, R.E. 1974. Vegetative propagation and the genetic improvement of North American hardwoods. In: 45 (p. 211-20).
28. Fielding, J.M. 1970. Trees grown from cuttings compared with trees grown from seed (*Pinus radiata* D. Don.). Silv. Gen. 19:54-63.
29. Fortanier, E.J. and H. Jonkers. 1976. Juvenility and maturation of plants as influenced by their ontogenetical and physiological ageing. Acta Hort. 56:37-44.
30. Franclet, A. 1980. Rajeunissement et propagation végétative des ligneux. 1980 AFOCEL Annales de Recherches Sylviloles, 315 p. (p. 11-42).
31. Gharyal, P.K. and S.C. Maheshwari. 1981. In vitro differientation of somatic embryoids in a leguminous tree - *Albizzia lebbeck* L. Naturwissenschaften, 68:379-80.
32. Gill, J.G.S. 1983. Comparison of production costs and genetic benefits of transplants and rooted cuttings of *Picea sitchensis*. Forestry, 56(1):61-73.
33. Griffin, A.R., J.N. Cameron and F.C. Hand. 1976. Production of radiata pine cuttings for seed orchard establishment. Aust. For. 39(2):121-30.
34. Gupta, P.K., A.F. Mascarenhas and V. Jagannathan. 1981. Tissue culture of forest trees - clonal propagation of mature trees of *Eucalyptus citriodora* Hook, by tissue culture. Pl. Sci. Letters, 20:195-201.
35. Gupta, P.K., A.L. Nadgir, A.F. Mascarenhas and V. Jagannathan. 1980. Tissue culture of forest trees: clonal multiplication of *Tectona grandis* L. (teak) by tissue culture. Pl. Sci. Letters, 17:259-68.

24

36. Haissig, B.E. 1970. Preformed adventitious root initiation in brittle willows grown in a controlled environment. Can. J. Bot. 48:2309-12.

37. Haissig, B.E. 1982. Activity of some glycolytic and pentose phosphate pathway enzymes during the development of adventitious roots. Physiol. Plant. 55:261-72.

38. Hare, R.C. 1975. Girdling promotes rooting of slash pine cuttings. In: Proc. of the 13th Southern Forest Tree Improvement Conference (p. 226-229).

39. Hartmann, H.T. and D.E. Kester. 1983. Plant propagation: principles and practices. 4th ed. Prentice-Hall, Inc., Englewood Cliffs, N.J., 727 p.

40. Hartney, V.J. 1980. Vegetative propagation of the eucalypts. Aust. For. Res. 10:191-211.

41. Hedstrom, B.S. and P. Krutzsch. 1982. Regulations on clonal forestry with Picea abies. In: 47 (p. 109-12).

42. Heybroek, H.M. 1978. Primary considerations: multiplication and genetic diversity. Unasylva, 30:27-33.

43. Hood, J.V. and W.J. Libby. 1978. Continuing effects of a maturational state in radiata pine and a general maturational model. In: K.W. Hughes, R. Henke, and M. Constantin, eds. Propagation of higher plants through tissue culture: a bridge between research and application April 16-19, 1978, USDOE - Univ. of Tennessee, Knoxville, Tenn., 305 p. Available from NTIS Springfield, VA.; Conf. - 7804111.

44. Hu, C.Y. and I.M. Sussex. 1971. In vitro development embryoids on cotyledons of Ilex aquifollum. Phytomorph. 21:103-7.

45. IUFRO. 1974. Proceedings of the International Forest Research Organisation's Working Party S2.01.5 on reproductive processes 12-16 November 1973 in Rotorua, New Zealand. N.Z. J. For. Sci. (Special issue on vegetative propagation) 4(2):119-458.

46. IUFRO. 1981. IUFRO International Workshop on in vitro cultivation of forest tree species. AFOCEL, Nangis, France, 363 p.

47. IUFRO. 1982. Proceedings of the IUFRO joint meeting of working parties on genetics about breeding strategies including multiclonal varieties. Lower Saxony Forest Research Institute, Escherode, West Germany, 238 p.

48. Jacquiot, C. 1955. Sur le role des correlations d'inhibition dan les phenomenes d'organogenese observes chez le tissu cambial, cultive in vitro de certain abres. Incidences sur les problems du bouturage. C.R. Acad. Sci. 241:1014-66.

49. Karnosky, D.F. 1981. Potential for forest tree improvement via tissue culture. Bioscience, 31(2):114-20.

50. Kemperman, J.A. and B.V. Barnes. 1976. Clone size in American aspens. Can. J. Bot. 54:2603-7.

51. Kirby, E.G. 1982. The use of in vitro techniques for genetic modification of forest trees. In: 13 (p. 369-386).

52. Klaehn, F.U. 1963. The relation of vegetative propaga-
    tion to topophysis, cyclophysis and periphysis in forest
    trees. In: Proc. 10th Northeastern Forest Tree Improve-
    ment Conference (p. 42-50).
53. Kleinschmit, J. 1974. A program for large-scale cutting
    propagation of Norway spruce. In: 45 (p. 359-66).
54. Kleinschmit, J. and J. Schmidt. 1977. Experiences with
    Picea abies cuttings propagation in Germany and problems
    connected with large scale application. Silv. Gen. 26
    (5-6):197-203.
55. Kochba, J., G. Ben-Hayyim, P. Spiegel-Roy, S. Saad, and
    H. Neuman. 1982. Selection of stable salt tolerant
    callus cell lines and embryos in Citrus sinensis and C.
    auranthium. Z. Pflanzenphysiol. 106:111-8.
56. Kochba, J. and P. Spiegel-Roy. 1977. The effect of
    auxin, cytokinins and inhibitors on embryogenesis in
    habituated ovular callus of the "Shamouti" orange
    (Citrus sinensis). Z. Pflanzenphysiol. 81:283-8.
57. Konar, R.N. and Y.P. Oberoi. 1965. In vitro development
    of embryoids on the cotyledons of Biota orientalis.
    Phytomorph. 15:137-40.
58. Lahiri, A.K. 1981. A note on layering of Cryptomeria
    japonica Don. and Pinus patula. Indian Forester, 107
    (1):51-3.
59. Lakshmi Sita, G., N.V. Raghava Ram and C.S. Vaidyanathan.
    1979. Differentiation of embryoids and plantlets from
    shoot callus of sandalwood. Pl. Sci. Letters, 15:265-70.
60. Lambeth, C.C. 1980. Juvenile-mature correlations in
    Pinaceae and implications for early selection. For. Sci.
    26:571-80.
61. Leach, G.N. 1979. Growth in soil of plantlets produced
    by tissue culture. Tappi, 62(4):59-61.
62. Légére, A. and S. Payette. 1981. Ecology of a black
    spruce (Picea mariana) clonal population in the hemiartic
    zone, Northern Quebec: population dynamics and spatial
    development. Arctic and Alpine Res. 13(3):261-76.
63. Lepistö, M. 1974. Successful propagation by cuttings
    of Picea abies in Finland. In: 45(p. 367-70).
64. Libby, W.J. 1982. What is the safe number of clones per
    plantation. In: H.M. Heybroek, B.R. Stephan and K. von
    Weissenberg, eds., Resistance to Diseases and Pests in
    Forest Trees, Proceedings of the 3rd International Work-
    shop on the Genetics of Host-Parasite Interactions in
    Forestry, Center for Agricultural Publishing and Documen-
    tation, Wageningen, The Netherlands (p. 342-60).
65. Libby, W.J., A.G. Brown and J.M. Fielding. 1972. Effects
    of hedging radiata pine on production, rooting and early
    growth of cuttings. N.Z. J. For. Sci. 2(2):263-83.
66. Lowery, R. 1980. Production of Pinus caribaea var.
    Hondurensis planting stock using pregirdled stem cuttings.
    Malaysian Forester, 43:16-23.
67. Morgan, D.L. and E.L. McWilliams. 1976. Juvenility as
    a factor in propagating Quercus virginiana mill. Acta
    Hort. 56:263-7.

68. Mott, R.L. 1981. Trees. In: B.V. Conger, ed., Cloning agricultural plants via in vitro techniques, CRC Press, Boca Raton, Florida, 273 p. (p. 217-54).

69. Muhs, H.-J. 1982. Identifying problems and legal aspects of using multiclonal mixtures. In: 47 (p. 92-108).

70. Mullins, M.G., Y. Nair and P. Sampet. 1979. Rejuvenation in vitro: induction of juvenile characteristics in an adult clone of Vitis vinifera L. Ann Bot. 44:623-7.

71. Manson, A. 1982. Closing topics. In: 47 (p. 228-31).

72. Pence, V.C., P.M. Hasegawa and J. Janick. 1980. Initiation and development of asexual embryos of Theobroma cacao L. in vitro. Z. Pflanzenphysiol. 98:1-14.

73. Radojević, L. 1979. Somatic embryos and plantlets from callus cultures of Paulownia tormentosa Steud. Z. Pflanzenphysiol. 91:57-62.

74. Radojeic, L., R. Vujicic and M. Neskovic. 1974. Embryogenesis in tissue culture of Corylus avellana L. Z. Pflanzenphysiol. 77:31-41.

75. Rauter, R.M. 1982. Recent advances in vegetative propagation including biological and economic considerations and future potential. In: 47 (p. 33-57).

76. Rauter, R.M. 1983. Clonal forestry: its impact on tree improvement and our future forests. Proceedings of the 19th Biennial Meeting of the Canadian Tree Improvement Association, Toronto, Ontario, Canada, August 23-26, 1983, (in press).

77. Roberts, A.N. and F.W. Moeller. 1978. Speculations relating the loss of rooting potential with aging of Douglas-fir stock plants. The Plant Propagator, 24(1): 2-4.

78. Romberger, J.A. 1976. An appraisal of prospects for research on juvenility in woody perennials. Acta Hort. 56:301-17.

79. Rook, D.A. and J.F.F. Hobbs. 1975. Soil temperatures and growth of rooted cuttings of radiata pine. N.Z. J. For. Sci. 5(3):296-305.

80. Ross, S.D. 1975. Production, propagation and shoot elongation of cuttings from sheared 1-year-old Douglas-fir seedlings. For. Sci. 21:298-300.

81. Roulund, H. 1975. The effect of the cyclophysis and topophysis on the rooting and behavior of Norway spruce cuttings. Acta Hort. 54:39-50.

82. Roulund, H. 1978. A comparison of seedlings and clonal cuttings of Sitka spruce (Picea sitchensis (Bong.) Carr.). Silv. Gen. 27(3-4):104-8.

83. Roulund, H. 1981. Problems of clonal forestry in spruce and their influence on breeding strategy. For. Abst. 42(10):457-71.

84. Saito, A. 1980. In vitro differentiation of embryoids from somatic callus tissues in Aesculus. J. Jap. For. Soc. 63(8):308-10.

85. Sakai, A. and Y. Sugawara. 1975. Survival of poplar callus at super-low temperatures after cold acclimation. Pl. Cell Physiol. 14:1201-4.

86. Sato, T. 1974. Callus induction and organ differentation in anther culture of poplars. J. Jap. For. Soc. 56:55-62.
87. Schier, George A. 1981. Physiological research on adventitious shoot development in aspen roots. USDA Forest Service Intermountain Forest and Range Experiment Station General Technical Report INT-107, 12 p.
88. Schreiner, E.J. 1939. The possibilities of the clone in forestry. J. For. 37:61-2.
89. Shelbourne, C.J.A. and I.J. Thulin. 1974. Early results from a clonal selection and testing program with radiata pine. In: 45 (p. 287-98).
90. Sheppard, J.F., D. Bidney and E. Shahin. 1980. Potato protoplasts in crop development. Science, 208:17-24.
91. Skrøppa, T. 1981. Some results from a 20-year-old cutting experiment with Norway spruce. In: 26 (p. 105-15).
92. Smith, D.R., J. Aitken and G.B. Sweet. 1981. Vegetative amplification - an aid to optimising the attainment of genetic gains from Pinus radiata. In: S.L. Krugman and M. Katsuta, eds., Proceedings of the Symposium on Flowering Physiology at the XVII IUFRO World Congress, 141 p. (p. 117-23).
93. Smith, D.R., K.J. Horgan and J. Aitken-Christie. 1982. Micropropagation of Pinus radiata for afforestation. In: A. Fujiwara, ed., Plant Tissue Culture 1983, Maruzen Co. Ltd., Tokyo, 839 p. (p. 723-4).
94. Sommer, H.E. and C.L. Brown. 1980. Embryogenesis in tissue cultures of sweetgum. For. Sci. 20(2):257-60.
95. Staritsky, G. 1970. Embryoid formation in callus tissue of coffee. Acta Bot. Neerl. 19(4):509-14.
96. Steinhauer, A. 1981. Haploid tissue culture for hybrid breeding in forestry. In: 46 (p. 295-306).
97. Stoltz, L.P. and C.E. Hess. 1966. The effect of girdling on root initiation: carbohydrates and amino acids. Proc. Am. Soc. Hort. Sci. 89:734-42.
98. Sugawara, Y. and A. Sakai. 1974. Survival of suspension cultured sycamore cells cooled to the temperature of liquid nitrogen. Pl. Physiol. 57:722-24.
99. Sweet, G.B. and L.G. Wells. 1974. Comparison of the growth of vegetative propagules and seedlings of Pinus radiata. In: 45 (p. 399-409).
100. Thompson, D.G. and J.B. Zaerr. 1981. Induction of adventitious buds on cultured shoot tips of Douglas-fir (Pseudotsuga menziesii (Mirb.) Franco). In: 46 (p. 166-174).
101. Thulin, I.J. and T. Faulds. 1968. The use of cuttings in the breeding and afforestation of Pinus radiata. N.Z. J. For. 13:66-77.
102. Tisserat, B. 1979. Propagation of date palm (Phoenix dactylifera L.) in vitro. J. Exp. Bot. 30:1275-83.
103. Toda, R. 1974. Vegetative propagation in relation to Japanese forest tree improvement. In: 45 (p. 410-17).
104. Ward, W.W. 1978. Knowing the forest - sprouts, suckers and epicormics. National Woodland, 1(10):5-7.

105. Wells, L.G. 1974. Differential browsing of <u>Pinus</u>
      <u>radiata</u> cuttings, grafts and seedlings. N.Z. J. For.
      19(1):138-40.
106. Werner, M. 1977. Vegetative propagation by cuttings of
      <u>Picea abies</u> in Sweden. <u>In</u>: 25 (p. 97-102).
107. Wilcox, M.D., I.J. Thulin and T.G. Vincent. 1976.
      Selection of <u>Pinus radiata</u> clones in New Zealand for
      planting from cuttings. N.Z. J. For. 21(2):239-47.
108. Wochok, Z. and M. Abo El-Nil. 1977. Conifer tissue
      culture. Proc. Int. Pl. Prop. Soc. 27:131-6.
109. Zobel, B. 1981. Vegetative propagation in forest
      management operations. <u>In</u>: Proceedings of the 16th
      Southern Forest Tree Improvement Conference. (p. 149-59).
110. Zobel, B., E. Campinhos and Y. Ikemori. 1983. Selecting
      and breeding for desirable wood. Tappi, 66(1):70-4.

2. PROPAGATION AND PRESERVATION OF ELMS VIA TISSUE CULTURE SYSTEMS

D. F. KARNOSKY AND R. A. MICKLER

Director, Center for Intensive Forestry in Northern Regions, Department of Forestry, Michigan Technological University, Houghton, Michigan 49916, and Research Associate, New York Botanical Garden Cary Arboretum, Box AB, Millbrook, New York 12545

ABSTRACT

Tissue culture systems offer opportunities to propagate, select preserve, and improve forest trees. This paper discusses the status of callus, suspension, anther, and protoplast cultures and cryopreservation in elms. These findings are related to broader uses in reforestation programs. The paper also describes some research needs in each of these areas.

2.1 INTRODUCTION

2.1.1 Importance of elms

For many centuries, the elms (Ulmus spp.) have been important components of forests, hedgerows, windbreaks, and urban areas in north temperature regions around the world. Elms have been used for over 2,000 years in Europe, Central Asia, and China for cattle feed. More recently they have been used for many different types of wood products, including boats, piles, furniture, veneer, paneling, flooring, containers, fence posts, and caskets.

The most conspicuous use of elms, however, has been in landscape plantings. Elms have been extensively used in urban and suburban landscape plantings, especially in Europe and North America.

American elm (Ulmus americana L.), perhaps the most popular of the more than 25 elm species and often referred to as one of nature's noblest vegetables, was planted very commonly along street sides and in parks across North America in the 1800's and early 1900's. Because of its elegant vase-shaped crown and high,

Duryea, M.L. and Brown, G.N. (eds.). Seedling physiology and reforestation succes
©1984, Martinus Nijhoff/Dr W. Junk Publishers, Dordrecht/Boston/London. ISBN 978-94-009-6139-5

arching branches which create a cathedral-like effect over
streets, American elm was often planted in near monocultural
fashion in large and small cities alike. This did not seem risky
at the time since American elm exhibited excellent urban
hardiness, being tolerant of de-icing salts, air pollution, and
soil compaction, and growing rapidly on a wide variety of sites
and soil types (7, 8).

2.1.2 Elm diseases

America's love affair with American elm came to an abrupt halt
in the middle of this century when the devastation of Dutch elm
disease became apparent. First identified as a problem in Europe
at the end of the First World War, Dutch elm disease, caused by
the vascular-wilt fungus Ceratocystis ulmi (Buisman) C. Moreau
started to kill trees in North America in the early 1930's.
During the some 50 years that the disease has been present in
North America, its range has steadily increased and it has now
killed an estimated 50 to 100 million elm trees from coast to
coast. Large cities such as Buffalo, Chicago, Milwaukee,
Minneapolis, and St. Paul have lost hundreds of thousands of
elms.

A second major disease problem of elms has begun to raise havoc
with the remaining elm populations in this country. For the past
twenty years, phloem necrosis has been increasing in importance in
the eastern United States. Also called "elm yellows", this
disease is caused by mycoplasma-like organisms. Together, Dutch
elm disease and phloem necrosis make any additional planting of
American elm in North America risky at best.

2.1.3 Tissue culture systems applied to elm improvement

Most elm improvement programs have concentrated on utilizing
disease resistance from Asian and European elm germplasm in various
hybrid combinations. Among the most promising elms for disease
resistance are two Asian elms, Chinese elm (U. parvifolia Jacq.)
and Siberian elm (U. pumila L.). Unfortunately, these resistant
elms do not have the crown shape, form, and public acceptance of
the American elm. Thus, a hybrid with the ornamental characteristics of American elm and the disease resistance of the Asian

elms became the goal of an elm breeding program begun about ten years ago at the New York Botanical Garden Cary Arboretum. This was no small challenge, considering that no one had ever successfully crossed the tetraploid (2n=56 chromosomes) American elm with any other elm species (all of which are diploid with 2n=28 chromosomes).

The potential applicability of tissue culture systems for anther, callus, suspension cell, and protoplast culture and for cryopreservation of selected individuals became apparent in this elm improvement program. The remainder of this paper describes the state-of-the-art of elm tissue culture and explores the applicability of the various tissue culture systems to reforestation programs. Additional research needs are also discussed.

2.2 CALLUS AND SUSPENSION CULTURES

Elms were among the first tree species used in tissue culture experiments. Gautheret (4, 5) attempted to define the cultural conditions for shoot formation and development from excised cambial tissue of Ulmus campestris L. by studying the effects of different sugars, light wavelength and energy, vitamin B_1 and auxins. The first major work with callus and suspension cultures with elms did not occur, however, until some 30 years later when Chalupa (1) reported on the effects of hormones on organogenesis in U. campestris callus cultures. Durzan and Lopushanski (3) in the same year reported on their successful regeneration of whole plants from callus derived from suspension cultures. They recognized that compared to standard propagation methods in forestry, tissue culture offered an attractive system for vegetative propagation of superior trees.

Subsequent advances in callus culture techniques with elms included more rapid and dependable callus growth and a substantial reduction in time and increased dependability of in vitro propagation of American elm as reported by Lange (10) and Karnosky, Mickler, and Lange (9).

We presently use the media described by Kao (6) for initiation and maintenance of callus and suspension cultures. For organo-genesis of the callus cultures, however, we use the woody plant medium described by Lloyd and McCown (12). Our shoot induction procedure involves the supplementing of the woody plant medium (in an agar form) with 4.0 mg/l 6-benzylamino purine (BAP) and 0.1 mg/l a-napthalene acetic acid (NAA). Roots are then induced on shoots excised from the callus and placed on filter paper wicks immersed in the liquid woody plant medium supplemented with 1.0 mg/l a-napthalene acetic acid (NAA) and 0.1 mg/l indole-3-butyric acid (IBA). The total time from seed germination to callus formation to organogenesis to plantlet with shoots and roots can be as short as two months or less.

To date, we have been able to initiate 1 to 14 meristems from callus on each 1 cm length of hypocotyl in approximately 90% of the random genotypes of American elm that we have tested. Following further shoot development and subsequent axillary bud break, the numbers of shoots can be rapidly increased. These shoots are readily rooted as previously described. By using our techniques in a cycle of seed germination → callus initiation→ suspension cell culture with biweekly subculture → callus cultures → organogenesis, we feel that many thousands of plantlets per genotype could be produced each year.

Elms are still one of the few tree species from which whole plants have been regenerated from callus. Thus, this remains a very important area of research, especially in the conifers where no one has yet obtained whole plants from callus.

Among the many beneficial uses of clonally propagated plantlets from a tissue culture program in forest tree improvement are:

1. Establishment of seed orchards and/or seed production areas;
2. Evaluation of genetic material on a wide variety of sites and in different climatic conditions (genotype-environment studies); and
3. Mass propagation of superior trees for reforestation programs.

Another important area of future research with elms and with many other tree species is the induction of embryogenesis from suspension cell cultures. Embryogenesis offers the potential for greater propagation efficiency than does organogenesis. We suspect that the cultural conditions for embryogenesis induction are dependent on developing the proper hormonal regimen in combination with a liquid medium such as the woody plant medium.

Another area of tissue culture needing additional research is the development of improved systems for shoot formation and development from callus cultures of mature trees. Chalupa (2) tested in vitro propagation of mature trees of U. campestris from axillary buds. Multiple shoots were produced from axillary buds on the original shoots and from the callused bases of shoot segments.

2.3 ANTHER CULTURE

Since American elm is the only tetraploid elm, halving its chromosome number would facilitate breeding at the 2n=28 chromosome diploid level. This would avoid the problem of sterile triploids being produced in diploid/tetraploid crosses. It would also simplify segregation patterns in hybrids, and perhaps be advantageous in disease resistance if the vessel sizes of dihaploids were reduced, as might be expected. Smaller vessels would be less conducive to Ceratosystis ulmi hyphae growth than those of the tetraploid.

Redenbaugh, Westfall, and Karnosky (15) cultured the first verified dihaploid callus of American elm. Best results were obtained on media with a high (3% or greater) sucrose concentration and a high concentration of auxin or auxin with cytokinin. Tree-to-tree differences and variation in the stage of microspore development were important in the amount of dihaploid callus production. Incubation under dark conditions of anthers after tetrad separation but before the first mitosis produced best results. While budlike structures were formed on the callus, no further development occurred.

This area of research remains a very important one, not only for elms but for other tree species as well. A potential use of haploids is the production of isogenic lines. Use of these lines in heterosis breeding offers the potential of improving species which show little variability and/or that have a high degree of inbreeding depression or self-incompatibility. Haploid plants and/or cell cultures may also prove valuable in isolating gene locations and in studying inheritance patterns in tree species. Basic studies are needed to further characterize media, hormone, and culture conditions to get a proliferation of American elm dihaploid callus and to determine how stable these cultures are in terms of chromosome numbers.

2.4 PROTOPLAST CULTURE

Elm protoplasts have been isolated from pollen mother cells, tetrads, microspores, cotyledons, and callus and suspension cultures (10, 11, 13, 14). We have obtained cell wall regeneration after 4 to 21 days in culture and cell division after 9 to 21 days. The farthest these cultures have progressed are to the stage of multi-cellular, torpedo-shaped embryoids. Thus, there are still numerous basic experiments that need to be done to use protoplast systems for elm improvement. The most critical of these appear to be: 1. the selection of fast-working enzymes with reduced toxicity for releasing the protoplasts; and 2. the identification of suitable osmotic conditions, culture medium, and hormone treatments for protoplast culture.

The potential uses of protoplasts for cell fusion, induction of somaclonal variation, and the incorporation of plasmids or organelles into cells are all exciting possibilities for the future in forest tree improvement. To date, however, no plants from any major forest tree have been regenerated from protoplasts.

2.5 CRYOPRESERVATION

In a cooperative effort between our laboratory and the USDA Cryogenic Laboratory at Berkeley, we have been able to demonstrate that American elm callus can survive liquid nitrogen immersion (17) Callus pieces were treated with a cryoprotectant mixture of polyethylene glycol, glucose, and dimethylsulfoxide (10%, 8%, and 10% w/v), frozen at $1^{o}C$ per minute to $-30^{o}C$ and then immersed in liquid nitrogen. Plantlets were regenerated from thawed callus. This demonstrates that short-term preservation of American elm germplasm is now possible via cryopreservation. Additional studies with callus and established vegetative materials are needed to determine if long-term storage of elm germplasm is possible.

Cryopreservation offers an opportunity to efficiently and safely preserve germplasm of forest tree species that are being lost due to insect and/or disease epidemics. However, its greatest use may come in traditional tree improvement programs where large amounts of space and time are needed to evaluate genetic studies. With conifers, for example, where the ability to clonally propagate plants via tissue culture systems requires juvenile plant material but where long-term studies are needed to evaluate growth rates, form, cold hardiness, and wood quality, the cryopreservation of cotyledons or of callus on cotyledons may provide a mechanism to preserve all genotypes of a given genetic experiment until the field test is completed and then propagate only those that have been shown to be genetically superior. The preservation of germplasm for future generations of tree breeding is also important, especially for species with small natural ranges.

Studies with other tree species and for longer storage periods will be necessary to develop the true potential of cryopreservation. The concentrations and types of cryoprotectants, the rate of freezing, and the condition of the samples to be frozen will all have to be examined on a species-by-species basis.

2.6 SUMMARY

Significant progress has been made in the past ten years in elm tissue culture systems. It is now possible to induce organogenesis from many different types of cell and tissue systems with several elm species. Furthermore, it is possible to culture plants from single cell suspension cultures of American elm, and we can use cryopreservation techniques to store American elm callus in liquid nitrogen. Additional research is needed for. further use of tissue culture systems in elm improvement programs. Studies of anther culture, protoplast culture, embryogenesis, and disease resistance testing are needed and offer great potential for the future.

36

REFERENCES

1. Chalupa, V. 1975. Induction of organogenesis in forest tree cultures. Commun. Inst. For. Czech. 9:39-50.
2. Chalupa, V. 1979. In vitro propagation of some broad-leaved forest trees. Commun. Inst. For. Czech. 11:159-170.
3. Durzan, D. J. and S. M. Lopushanski. 1975. Propagation of American elm via cell suspension cultures. Can. J. For. Res. 5:273-277.
4. Gautheret, R. J. 1940a. Recherches sur le bourgeonnement du tissu cambial d'Ulmus campestris cultivé in vitro. C. R. Acad. Sci. 210:632-634.
5. Gautheret, R. J. 1940b. Nouvelles recherches sur le bourgeonnement du tissu cambial d'Ulmus campestris cultivé in vitro. C. R. Acad. Sci. 210:744-746.
6. Kao, K. N. 1977. Chromosomal behavior in somatic hybrids of soybean - Nicotiana glauca. Molec. Gen. Genet. 150:225-230.
7. Karnosky, D. F. 1979. Dutch elm disease: A review of the history, environmental implications, control, and research needs. Environ. Conserv. 6:311-321.
8. Karnosky, D. F. 1982. Double jeopardy for elms: Dutch elm disease and phloem necrosis. Arnoldia. 42:70-77.
9. Karnosky, D. F., R. A. Mickler, and D. D. Lange. 1982. Hormonal control of shoot and root induction in hypocotyl callus cultures of American elm. In Vitro. 18:275. (Abstract).
10. Lange, D. D. 1981. Techniques for isolating, purifying, and culturing Ulmus protoplasts. MS Thesis, State University of New York, College of Environmental Sciences and Forestry, Syracuse, New York. pp. 1-54.
11. Lange, D. D. and D. F. Karnosky. 1981. A discontinuous density gradient technique for purifying elm protoplasts. In Vitro. 17:228. (Abstract).
12. Lloyd, G. and B. McCown. 1980. Commercially feasible micropropagation of mountain laurel, Kalmia latifolia, by use of shoot-tip culture. Proc. Int. Plant Prop. Soc. 30:421-427.
13. Redenbaugh, M. K., D. F. Karnosky, and R. D. Westfall. 1981. Protoplast isolation and fusion in three Ulmus species. Can. J. Bot. 59:1436-1443.
14. Redenbaugh, M. K., R. D. Westfall, and D. F. Karnosky. 1980. Protoplast isolation from Ulmus americana L. pollen mother cells, tetrads, and microspores. Can. J. For. Res. 10:284-289.
15. Redenbaugh, M. K., R. D. Westfall, and D. F. Karnosky. 1981. Dihaploid callus production from Ulmus americana anthers. Bot. Gaz. 142:19-26.
16. Takai, S. 1974. Pathogenicity and cerato-ulmin production in Ceratocystis ulmi. Nature. 252:124-126.
17. Ulrich, J. M., R. A. Mickler, B. J. Finkle, and D. F. Karnosky. 1983. Survival and shoot differentiation in callus cultures of American elm after being frozen in liquid nitrogen. (In Preparation).

3. NEW FORESTS FROM BETTER SEEDS: THE ROLE OF SEED PHYSIOLOGY

F. T. BONNER

Plant Physiologist, USDA Forest Service, Southern Forest
Experiment Station, Post Office Box 906, Starkville,
Mississippi 39759

ABSTRACT

Full utilization of seed supplies requires a commitment
to improving tree seed technology, with the help of recent
advances in seed physiology. Studies of seed maturation
have led to new collection and extraction technology which
should increase seed yields. Computer analyses of test data
allow extension of seed testing beyond simple germination
percents to estimates of seed vigor, that elusive quality
factor which enables a seed to survive and grow. Basic
studies of the germination environment have revealed new
information on how temperature, imbibition, and seedcoat
microorganisms can be used to obtain vigorous, uniform
germination in the nursery bed or container.

3.1 INTRODUCTION

The time is fast approaching when practically all planting
stock of the major timber species will be produced from
genetically superior seeds. Better seeds call for better
technology for collecting, conditioning, testing, storing,
and planting. Advances in seed technology are also needed
for the numerous species used for other purposes than timber,
such as fuelwood, shelterbelts, agroforestry, and watersheds.
In many parts of the world, these species are of much more
critical importance than those grown solely for timber
production. Regardless of the intended use of the seedling,
better seed will aid reforestation, and knowledge of seed
physiology can be important to success. This paper will review
recent developments related to four stages of seed technology:

Duryea, M.L. and Brown, G.N. (eds.). Seedling physiology and reforestation succes
©1984, Martinus Nijhoff/Dr W. Junk Publishers, Dordrecht/Boston/London. ISBN 978-94-009-6139-5

(1) Collection, Cone Storage, and Conditioning; (2) Testing;
(3) Storage; and (4) Germination.

A detailed review of the biochemistry of tree seeds is
not attempted, although seed biochemistry is highly important.
The omission emphasizes the broad scope and complexity of the
subject. Some excellent recent books on this topic (9, 10, 42)
are available.

3.2 COLLECTION, CONE STORAGE, AND CONDITIONING

3.2.1 Collection

In spite of many years of study and practice, collectors
are still bringing in too many immature seeds. Good maturity
indices have been developed for most major species (15, 28),
but even when they are used conscientiously, optimum results
are not always obtained. One reason is the extreme natural
variation among trees in seed maturity dates (28). Logistical
constraints on the collection operation may force collectors
to take all seeds or cones from a particular locality at one
time, even though only a small portion may give positive
results in the maturity test.

Better information on seed maturation will help to solve
this problem. In northern Finland, where short growing
seasons often hinder full seed development, research on Scots
pine (57) has shown that variation in seed quality (maturity)
is more pronounced in poor seed years. Once the best trees
in poor years are identified, seeds from these trees can be
gathered in all average-to-poor crop conditions. The study
indicated that in good seed years, seeds from "best" trees
were no better than the average.

Another approach to this problem is to catch naturally
disseminated seeds on narrow mesh nets spread throughout the
orchard. The net collection system, developed by the Georgia
Forestry Commission 12 years ago (74), has proven successful
in Pinus taeda orchards in the South, and can surely be used
with many other species. Some physiological aspects of this
system are currently under study: What seed quality losses

occur when seeds are exposed to the weather for 2 to 3 months?
Is there a "natural stratification" which reduces need for
prechill before sowing? Is storage potential reduced?

3.2.2 Cone Storage

Cone lots of mixed maturity levels present special diffi-
cultuies in seed extraction. Immature cones generally contain
more moisture than fully mature cones, and the heat applied
in kilns to open the cones can reduce seed quality in the
greener cones. For this and other reasons, cones are usually
stored for a while in trays, boxes, or burlap bags to allow
slow drying before artificial heat is applied. Cone storage
also allows the final stages of maturation to take place in
some species, notably Abies (31).

Excessive moisture can be very dangerous if storage temperatures drop below freezing, not an uncommon condition
in the northern U.S. and Canada. Tanaka (64) reported that
the 50-percent-kill temperatures for seeds of Abies procera
at 20 and 35 percent moisture contents were -32° and -12°
respectively. Since seeds of this species require at least
a 1- to 2-month after-ripening period in the cones before
extraction (54), some protection is necessary.

Knowledge of seed behavior during cone storage is also
important for serotinous species. Pinus banksiana cones gave
acceptable yields and seed quality up to age 6 years past
maturity (3). After even one year, however, yields and
germination decreased, and abnormal germination increased.
The semiserotinous cones of Picea mariana do not do as well;
seed quality decreased rapidly in 2-year old cones before
stabilizing for 2 to 5 more years (34).

Artificial ripening, or cone storage designed purposely
to ripen immature seeds, has been proposed for at least 30
years as a way to extend the collection season, but it has
been very little used (28). Recent studies have shown that
proper cone handling can prevent harm to seeds of some species
in artificial ripening. Cones of Abies amabilis, Picea glauca,
P. sitchensis, P. englemannii, and Tsuga heterophylla were

kept in cone sheds or in refrigerated storage (2° C) for
6 months. The only loss in seed quality in either condition
could be traced to improper cone handling before storage (44).
Dry storage with good ventilation was essential. Haddon and
Winston (35) reported that Picea glauca seeds stored in cones
for 12 weeks at 5° C for artificial ripening still germinated
well 2 years after extraction. These authors also reported
that nonstratified seeds germinated better than stratified
seeds; this condition is common in weakened seeds.

3.2.3 Conditioning

Conditioning and upgrading of seed lots may seem to be a
matter of engineering rather than seed physiology, but one
new technique employs the hygroscopic properties of seeds to
separate living and dead seeds. By the IDS method, so called
by the Swedish scientists who developed it (61), Pinus contorta
seeds are imbibed for 16 hours, then incubated for 56 hours
at 15° C and 100 percent relative humidity. The seeds are
then dried for 12 hours at 15° C and 35 percent relative
humidity. During the drying, dead seeds lose almost all the
water previously imbibed, while living seeds retain most of
theirs. At this point, separation can be achieved by several
methods, such as water flotation, pneumatic separators, or
magnetic separators. Other species require conditioning at
different times and temperatures, but the key is to make the
separation at the point of greatest difference between dead
and live seed weights.

3.2.4 Application

Better knowledge of seed maturation and of the patterns of
variation has proved useful in seed orchard management. Cone
collection in many orchards is scheduled according to the
clonal ripening sequence, which can be established by careful
observation. Better seeds of many hardwood species are now
available to nurserymen because of improved maturity indices
for the seed collectors.

The IDS system of upgrading has great potential for solving one of the nurseryman's oldest problems. Elimination of dead seeds should lower container seed costs significantly and also lead to more uniform density in nursery beds.

3.3 TESTING

Seed physiologists have long been active in research on testing methods. Test procedures are needed not only for accurate measurement of all characteristics of seed lots, but also for evaluation of the response of test samples to various treatments. Three particular areas have received much attention in the last 15 years: X-ray radiography, mathematical modeling of germination, and vigor tests.

3.3.1 X-ray Radiography

The use of soft X-rays to examine seeds nondestructively is one development that tree seed workers did not adopt from agricultural research. This technique, developed with tree seeds in Sweden (52), has proved useful in many ways to identify empty seeds and damage from insects and disease, and mechanical causes. Examination by X-ray has its limitations, however; filled dead seeds usually cannot be distinguished from filled live seeds.

A handbook for basic radiographic interpretation of the major forest tree species has been published (7). Many sophisticated medical X-ray techniques can be adapted to seed radiography (66), but equipment costs or limited accessibility have prevented frequent use. Research in X-ray radiography techniques for tree seeds is minimal at present, but application of the earlier findings is common in many laboratories.

3.3.2 Mathematical Modeling of Germination

Seed scientists have always been intrigued by the possiblity of expressing the germination of a sample over time in mathematical terms. Quantitative expressions based on polynomial regressions (32), the normal distribution (40), probit transformations of cumulative germination percentages (21), the

logistic function (58), and the Weibull function (17) have
been proposed. Seeds from wild populations do not react
like the homogeneous varieties of non-dormant agricultural
seeds. Frequency distributions of tree seed germination are
almost always skewed positively, although negative skewness
is sometimes encountered. The Weibull function model is the
only one flexible enough to handle both types of skewness
(FIG. 1). Its principal drawback is that the biological
interpretation of some model parameters is not yet clear.

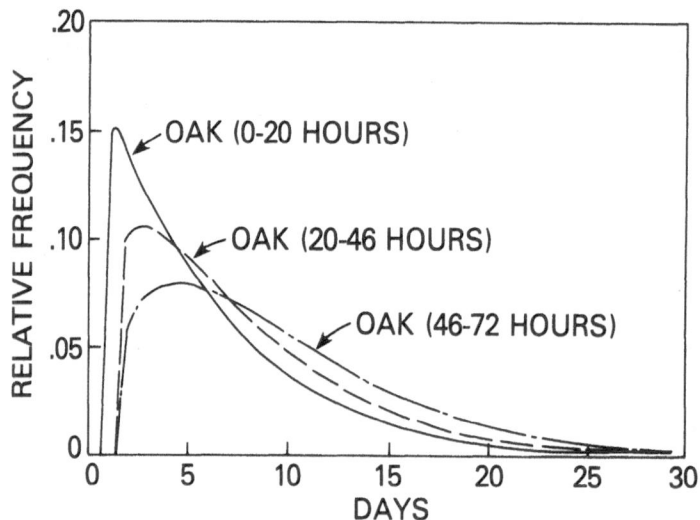

FIGURE 1. Relative germination frequency [f(x)] of three
    treatment groups of white oak, based on average Weibull
    parameters. Seeds received accelerated aging at 40° C
    for the times noted (17).

A new model for expressing germination frequency has been
proposed by Bould and Abrol (20). They suggest that the
observed cumulative distribution of time to germination in
a sample is a mixture of exponential distributions, each of
which can characterize a subgroup within the population.
The number of subgroups can be estimated and their distributions
calculated. This model is interesting because it deals with
the subpopulations (subgroups) we know to exist in heterogeneous
seed lots. Since practically all tree seed lots display this
heterogeneity, it will be interesting to see if this model

can be applied in tree seed research.

Each of these models of germination has worked well with certain seed lots and under certain test conditions; none seem to have universal application. But one point seems clear: the mathematical study of biological processes is progressing. The advent of the computer age will soon make complex routines such as fitting Weibull models commonplace in biological research laboratories. This capability must be used in tree seed physiology research.

3.3.3 Vigor Tests

The Association of Official Seed Analysts has defined seed vigor as "those seed properties which determine the potential for rapid uniform emergence and development of normal seedlings under a wide range of field conditions" (2). Development and standardization of tests for seed vigor has perhaps received more attention from seed scientists over the last decade than any other aspect of seed research. Note the distinction between vigor tests and standard laboratory germination tests. Germination tests measure germination potential under optimum conditions; the resulting percentage is the highest you should expect. Vigor tests seek to measure the ability of seeds to germinate rapidly and uniformly under a wide range of conditions, including stressful ones.

Vigor measurements can be classified as direct growth tests, indirect tests based on physiological measurements, and extrapolation from laboratory germination data. The first type, based on actual seedling growth in incubators or greenhouses, has not been tested extensively with trees, nor has it been widely used for vigor evaluation of crop seeds (39). The presence of genotype or seed source differences and dormancy decrease the attractiveness of these tests for tree seeds, because although the factors could reduce actual growth in a relatively short test period, they might not affect field performance in nurseries over an entire growing season.

Estimations of vigor levels from germination test data are popular because the measurements are familiar, and a second test is not required. The only extra effort comes from more frequent counting of germination than is normal and from additional data analyses. Czabator's Peak Value (PV) and Germination Value (GV) have been widely used (24). The PV is the maximum quotient derived from dividing cumulative germination percent on each day by the number of days of elapsed test time. The PV is considered a good germination rate parameter and has often been significantly correlated with seed performance (12, 17). The GV = MDG x PV, where mean daily germination (MDG) is total percent germination divided by total days in the test. The GV has also been used successfully in seed research as an index which combines both rate and completeness of germination (6, 27).

Some other useful germination parameters are mean time to germination (MGT), time required for 50 percent (or other proportions) of germination (G_{50}), and standard deviation of germination times (a measure of uniformity of germination). These parameters can be derived from germination test data or from models developed from the data, such as the Weibull model (17) or the probit transformation (21). The latter seems to give a good fit only if no dormancy is present, however.

Indirect estimates of seed vigor through some physiological measurement have received much study. Because most tests measure only one enzyme or one process, while germination results from the integration of many processes, it is not surprising that a suitable test has not been found. Tetrazolium salt (TZ) staining of tissue to measure dehydrogenase activity was one of the first techniques (49). Although TZ tests are still widely used as rapid viability estimates for dormant seeds (37), the method has not proved very useful as a standard vigor test (38).

Other biochemical tests proposed have been oxygen uptake by embryos (73), glutamic decarboxylase activity (33), and ATP activity (22). None of these have demonstrated acceptable consistency.

Two vigor tests measure an intact seed's reaction to controlled conditions: accelerated aging and electrical conductivity.

Both tests are very promising. In accelerated aging, seed
samples are subjected to high moisture and temperature (40°-45° C)
stress for short periods (usually up to 7 days), then germinated
in the usual manner. The amount of deterioration is reflected
in the germination response and is directly related to the
original vigor of the seeds. Although the deterioration of
seed tissues may not be the same as that which occurs in slow
aging in normal storage environments (10), the test has been
widely accepted for soybean vigor tests (2).

Experience with accelerated aging of tree seeds is limited.
Pitel (53) "aged" Pinus banksiana at 43° for up to 18 days.
As aging period increased, germination fell from 97 to 0 percent,
and seedling fresh weight decreased. Quercus rubra aged for
10 days had germination reduction of 100 to 42 percent. Similar
results have been obtained in the author's laboratory with
Platanus occidentalis, Fraxinus pennsylvanica, and Quercus
shumardii.

Pinus taeda and Liquidambar styraciflua show considerable
tolerance to accelerated aging, however, (Table 1). If initial
seed moisture content is as low as 8 percent, a short period
of aging will actually stimulate germination. As initial
moisture content is increased in L. styraciflua seeds, this
tolerance decreases. If heat only (40°), with no increased
moisture uptake, is applied to these species at 8 percent
moisture, germination does not decline until about 40 days for
P. taeda and 20 days for L. styraciflua. Germination of

Table 1. Germination at 28 days of seeds of two species which
received up to 4 days of accelerated aging (40°, 100% RH).
(Unpublished data, F. T. Bonner.)

| Original seed moisture (pct.) | Percent germination, by days aged | | | | | | | | | |
| | Pinus taeda | | | | | Liquidambar styraciflua | | | | |
	0	1	2	3	4	0	1	2	3	4
8	56	66	62	68	75	80	87	85	73	72
12	53	65	65	67	74	84	81	72	70	64
16	61	69	61	66	74	85	63	54	40	35
20	58	63	68	64	73	74	54	37	39	25

<u>P</u>. <u>taeda</u> was actually increased by this treatment up to 10 days. Although the relation between artificial aging and seed vigor has not been firmly established for any of these species yet, the potential may be there for a useful test.

In the electrical conductivity test, seed samples are leached in deionized water for specified periods at a constant temperature, and the conductivity of the leachate is measured. In theory, cell membranes are disrupted during seed deterioration, and electrolytes leach into the water and increase the conductivity. This technique is now widely used for many agricultural seeds (39), and promising results have been obtained for southern pines (19), <u>P</u>. <u>banksiana</u> (53), and <u>Quercus</u> <u>nigra</u> (1). Recent tests with <u>P</u>. <u>taeda</u> and <u>P</u>. <u>elliottii</u> produced significant correlation coefficients of 0.67 and 0.66 between mean conductivity reading and nursery bed emergence. The speed of the conductivity test makes it a very attractive method for tree seed managers.

3.3.4 Application

Successful modeling of seed germination has great potential for application through its use to quantify seed response to such conditions as seed source, environmental stress, seed treatment, deterioration in storage, and seed vigor. Refinement of modeling techniques to quantify responses from subpopulations of a seed sample can help explain strange treatment responses which sometimes baffle the analyst.

Seed vigor measurements can have wide application. As seeds age, vigor always declines before viability. Therefore, seed lots having lower vigor should be planted first and those with higher vigor stored for later years. In nurseries, seed lots with higher vigor should be sown first, since cold soil will affect these seeds the least. Seed lots with lower vigor should be planted later, when a more moderate bed environment will allow the best possible seedling emergence and establishment.

3.4 STORAGE

The two basic reasons for storing seeds are to prevent shortages for nursery sowing in poor seed years, and to conserve valuable germ plasm material. The seed manager's approach to storage will depend on the goal of his program. Seed storage problems now receiving most research attention are relevant to both goals: storage potential of orthodox seeds, recalcitrant seeds, and genetic damage during storage.

3.4.1 Storage Potential of Orthodox Seeds

Orthodox seeds can be dried to below 10 percent moisture and stored easily at subfreezing temperatures for long periods. Some orthodox genera are Pinus, Picea, Platanus, and Eucalyptus. Recalcitrant seeds must be maintained at high moisture contents (above 20 percent) and stored above freezing; they are very short-lived in storage. Quercus species make up the largest group of recalcitrant species in North America.

Orthodox seeds can be stored for long periods under ideal conditions. Under less than ideal conditions, these species might survive short-term storage (1 to 2 years) at little expense. Professor E. H. Roberts of England has proposed that seeds will always react in the same way under a given set of storage conditions (55). We can therefore derive survival models from storage data for each species, which can be used to predict storage performance at all conditions of storage. This theory has been tested with agricultural seeds (29) by the following equation:

$$v = K_i - p \, / \, 10^{K_E - C_w \log m - C_H t - C_Q t^2};$$

where v = probit percentage viability, p = storage period in months, m = seed moisture content (percent fresh weight), t = temperature (C°), K_i = probit percentage germination at zero storage time (a constant for the lot), and K_E, C_w, C_H, and C_Q are constants which apply to all genotypes and seed lots within a species.

The constants are determined from test data by fitting the following equation by multiple regression analysis:

$$\log \sigma = K_E - C_w \log m - C_H t - C_Q t^2,$$

where σ = the standard deviation of the distribution of seed "deaths" in time.

But do these survival models fit storage data on tree seeds? No experiments are complete as yet, but early results with four southern species (P. taeda, P. elliottii, Platanus occidentalis, and L. styraciflua) suggest that another model is needed. The probits of survival are not always linear over time (FIG. 2), a basic condition of the model (51). One possible alternative is to base the model on the Weibull function. Its flexibility seems to be very suitable for this purpose.

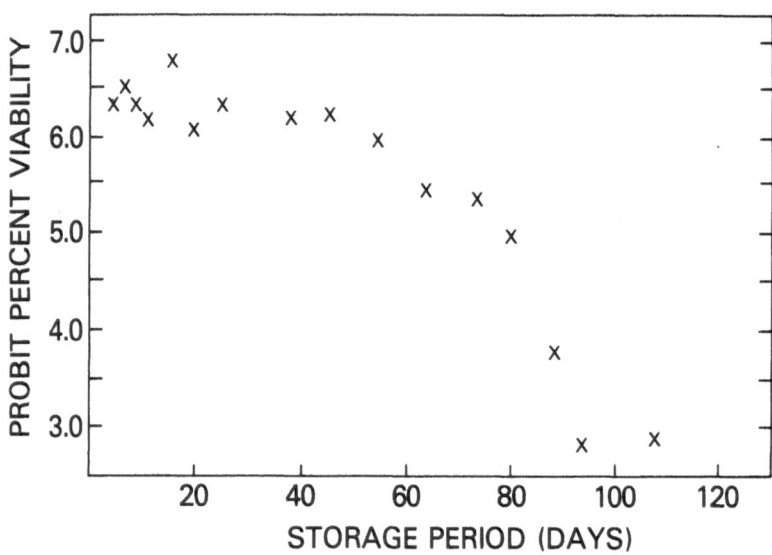

FIGURE 2. Probit transformation of percent germination over storage period for Pinus elliottii seeds stored at 40°, with 8 percent moisture content. (Unpublished data, F. T. Bonner.)

3.4.2 Recalcitrant Seeds

The greatest challenge in applied seed biology today is to find a method of long-term storage of recalcitrant seeds. Moderate success has been achieved with some Quercus species (subgenus Erythrobalanus, red oaks) by storage at 35 to 45 percent seed mositure and -1° to +3° (11, 63). At these high moisture contents, respiration proceeds at a rapid pace, and completely air-tight storage will kill the seeds. Some gas exchange must take place, but the seeds must not be allowed to dry out. Polyethylene bags with a wall thickness of about 0.1 mm provide a suitable storage environment.

Other Quercus species (subgenus Leucobalanus, white oaks) do not store as easily. These acorns, being low in fat and high in carbohydrates (14), can seldom be stored more than 6 months. Species with high acorn fat contents seem to store better, but the reason is unknown. One exception is Q. lyrata; this species has been stored for almost two years with very little loss in viability.

Many tropical species produce large seeds with high moisture contents, and most of these are recalcitrant in nature. Some are even more difficult to store than Quercus, however, because temperatures near freezing damage the seeds (43). Storage temperatures above 12° are necessary to prevent chilling injury.

The answer to the problem of recalcitrant seed storage may come from basic studies of cellular structure and chemistry. Recent work on seeds of Avicennia marina (8) suggests that metabolic activities during initial dehydration are very similar to metabolism typically associated with germination. Further dehydration disrupts subcellular processes to such a degree that death results, as it could in an orthodox seed in the stages of early germination. The tendency of most Quercus species to germinate while in storage supports this interpretation, as does a comparison of food reserve changes in southern species during storage (23) and during stratification and germination (67, 69).

3.4.3 Genetic Damage

As concern for the conservation of germ plasm has risen in recent years, so has our concern about possible genetic changes in seeds during storage. As some seeds age, even in proper storage, chromosome damage occurs and accumulates (56). Chromosome damage during storage has been reported for Fraxinus americana (65) and Pinus sylvestris (59), but evidence of real genetic changes in tree seeds is lacking. In fact, most studies on this problem have shown that chromosome aberrations which occur during seed aging are largely eliminated during early growth of the plant and are not transmitted to the next generation. Point mutations may occur, however, which will not be detected until the following generation (56). More research is definitely needed on this problem.

3.4.4 Application

Storage potential for tree species will be determined from models of seed survival. This information will be useful in planning storage facilities and managing seed stocks. For example, a manager who will keep a pine seed lot only one year before sowing does not need subfreezing storage. Storage potential data can tell him minimum conditions of temperature and seed moisture required for one year's storage without loss of seed quality.

Storage of recalcitrant seeds will not be a major problem in North America until we want to grow large quantities of oak seedlings. In tropical and subtropical areas, however, recalcitrant species are much more important. Solution of this problem could greatly stimulate tropical reforestation programs.

The importance of genetic damage to tree seeds in long-term storage is still not fully known. The potential impact of this damage to tree improvement and breeding programs, however, makes it imperative that we find the answers.

3.5 GERMINATION

The payoff for seed handling in a reforestation program is germination. We may have collected, tested, and stored our seed properly, but the acceptable final result is rapid, uniform, and vigorous germination. Although this aspect of seed technology may seem to be the easy part (why not just let the seeds do what comes naturally?), several important considerations deserve attention in research programs: environmental stress, microorganisms, and seed pretreatments.

3.5.1 Environmental Stress

Temperature and moisture are the principal stress agents for consideration in nursery beds. It has long been known that moisture stress reduces germination of seeds. Laboratory studies with osmotically-induced moisture stress have sought to define levels that limit germination (4, 18, 30, 50), but such conditions may not simulate natural environments at all. Other influences, such as temperature and light, can modify the moisture effect, and some results have shown that geographic (50), and clonal origin (30) can affect response to moisture stress.

The osmotic agents used in the laboratory studies can also be used in a procedure called osmotic priming to speed germination. In this technique, seeds are soaked in solutions of polyethylene glycol, which allows slow imbibition of water, but not enough for germination. Low-vigor seeds are helped, because rehydration occurs slowly, and thus does not physically damage the seed. In high-vigor seeds, moisture levels are increased to the point where little additional moisture is needed, and emergence may be quicker. Improved germination rates from PEG priming have been reported for Pinus patula (25) and P. sylvestris (60).

Temperature deserves more study. Barnett (6) has reported that Pinus taeda germination and seedling development is affected by the temperature at which water absorption occurs. Other studies with P. taeda have shown that alternating daily temperatures are better than constant temperatures (16, 26), although stratification improved performance at constant temperatures and cooler regimes. Optimum day and night temperatures were found to have a wide range, but a daily amplitude of 10° to 12° was best. Similar responses have been noted in many other plant species (45).

3.5.2 Microorganisms

Major impacts of microorganisms on seeds occur through pathogens and mycorrhizae. Full treatment of both topics is beyond the scope of this paper, and they are worthy of separate

consideration. Some recent findings, however, deserve mention
for their potential impact on seed management.

One significant report described two pathogenic fungi,
Diplodia gossypina and Fusarium moniliforme var. subglutinans,
which were isolated from both gametophyte and embryo tissues
of Pinus elliottii (48). The spread of pathogenic fungi by
way of seeds is certainly not a new discovery, but these results
with seed orchard material indicate a potential problem that
had not been recognized previously.

We should also note the recent discovery of highly branched
tubular channels through the pericarps of Quercus nigra, which
connect with pores in the exocarp (68). Not only does this
suggest a system for translocation of gases and moisture through
sclerified tissue to the embryo, but development of hypa-like
strands on the exocarp surface near the pores during stratification and germination suggests a possible method of entry
into the seeds by microorganisms.

The significance of the Q. nigra results is still unknown,
but it is evident that we know too little about the microenvir-
onment inside the seed coat. Do microorganisms enter the
Quercus acorns through the pores and tubular channels? Are
they pathogens, or are they beneficial fungi capable of forming
mycorrhizae? A water-soluble fungal inhibitor has been reported
in the pericarps of Acer saccharum fruits (71), but its exact
function is unknown. If the inhibitor leaches into the soil
around the seed, it might inhibit mycorrhizal fungi. This
entire area needs additional research.

3.5.3 Pretreatments

The standard pretreatment to overcome dormancy in seeds of
North American species has been cold, moist stratification at
2° to 5° C. One prescription per species has been the rule,
although variable degrees of stratification according to geographic source have been used for some species. But now, to
improve production efficiency, nurserymen are looking for ways
to get more rapid, uniform germination of each seedlot. As
germination rate increases, uniformity of emergence should

increase; as uniformity of emergence increases, seedling cull percent should decrease (70). It has also been estimated that the leading cause of seedling loss in southern nurseries is hard rains just after sowing.[1] Rapid, uniform germination can reduce the period in which seeds and young seedlings will be damaged.

One simple approach to improving nursery germination rate is to lengthen stratification to two or three times what is normally used. Numerous studies have shown that longer than normal stratification periods will increase germination rates only moderately under laboratory test conditions (27, 47, 72). When cooler test temperatures, which more closely match seedbed conditions, have been employed, however, a much greater benefit from long stratification treatments is evident (46). This technique may work well for high-vigor seed lots, but caution is needed. Extremely long pretreatment periods could have an opposite effect on low-vigor lots. Other approaches to pre-treatment such as aerated water soaks (5) have been tested, but there have been no basic changes in stratification tech-niques in a long time.

The use of chemicals as pretreatments to speed germination has attracted attention from time to time. Many species respond with faster germination rates to soaking in solutions of gibberellins (13), but little, if any, operational use of this method has been made. Increased rates of germination have been reported for hydrogen peroxide (62), citric acid (41), nitric acid (36), and seaweed concentrate (25), but like the gibberellin treatment, these are not used operationally. The cost of treating large quantities of seeds is a definite factor, and any successful chemical treatment will have to be either very cheap or extremely effective in increasing nursery emergence rate.

3.5.4 Application

The application of results from germination research should be very beneficial to nurseries. Improved pretreatments can

---

[1] David South, Auburn University, personal communication.

modify environmental stresses and lead to more rapid and uniform
germination, which will in turn decrease seedling cull percent
and increase the efficiency of seed utilization. The end
result is lower cost per plantable seedling, a goal of every
reforestation plan.

3.6 LOOKING AHEAD

How can research in seed physiology best serve forestry in
the years ahead? What are the research areas with the greatest
potential for payoff? Five problems seem to stand out:

Storage of recalcitrant seeds. This is the greatest challenge. Dormancy must be induced where nature provides none.
The greatest impact will be in tropical and subtropical
regions, but temperate-zone forestry will benefit also.

Rapid tests of seed quality. Often a 60-day test means
too long to wait to know the quality of the lot. The electrical conductivity test offers great potential, but other
ways are needed.

Chemical identification of seed origin. Test procedures
for varietal identification of crop species are becoming
very good. We should be able to borrow and adapt their
techniques for general seed source identification.

Population biology of seed lots. Different subpopulations
within a lot must be identified and characterized. Once
this is done, separation and custom treatment of subpopulations may be possible. For example, removal of the
dormant fraction of a seed lot would be very helpful.

Germination environments. Simple one-factor relationships
(temperature, moisture, pH, etc.) are not enough. The
dynamic interrelationships of all conditions, both inside
the seed and in its immediate surroundings, can be modeled.
With this information, environments can be altered (seed
coatings, soil additives, etc.) to provide rapid, uniform
germination and vigorous seedlings. This information could
also remove the mystery from some unexplained failures.

This is a short list, and many other worthy projects could
be added to it. I hope it will encourage more research in
tree seed physiology for the new forests of our future.

REFERENCES

1. Agmata, A. L. 1982. Physiological effects of drying Quercus nigra L. acorns. Unpubl. MS Thesis. Mississippi State University, Starkville, Mississippi. 59 p.
2. Association of Official Seed Analysts. 1983. Seed vigor testing handbook. A.O.S.A. Handb. No. 32. 88 p.
3. Baker, W. D. 1980. Natural seed storage in serotinous cones of Pinus banksiana. In IUFRO Int. Symp. on Forest Tree Seed Storage. Petawawa, Canada. p. 81-90.
4. Barnett, J. P. 1969. Moisture stress affects germination of longleaf and slash pine seeds. For. Sci. 15:275-276.
5. _____. 1971. Aerated water soaks stimulate germination of southern pine seeds. USDA For. Serv. Res. Pap. SO-67. 9 p.
6. _____. 1981. Imbibition temperatures affect seed moisture uptake, germination, and seedling development. In Proc. 1st South. Silv. Res. Conf. USDA For. Serv. Gen. Tech. Rpt. SO-34. p. 41-45.
7. Belcher, E. and J. A. Vozzo. 1979. Radiographic analysis of agricultural and forest tree seeds. Assoc. Off. Seed Anal. Handb. No. 31. 29 p.
8. Berjak, P., M. Dini, and N. W. Pammenter. 1983. Possible mechanisms underlying the differing dehydration responses in recalcitrant and orthodox seeds: desiccation-associated subcellular changes in propagulaes of Avicennia marina. 20th Int. Seed Test. Assoc. Congress, Ottawa, Canada. Preprint No. 33. 14 p.
9. Bewley, J. D. and M. Black. 1978. Physiology and biochemistry of seeds. 1. Development, germination and growth. Springer-Verlag, Berlin. 306 p.
10. _____. 1982. Physiology and biochemistry of seeds. 2. Viability, dormancy, and environmental control. Springer-Verlag, Berlin. 375 p.
11. Bonner, F. T. 1973. Storing red oak acorns. USDA For. Serv. Tree Plant. Notes 24. 3:12-13.
12. _____. 1974. Tests for vigor in cherrybark oak acorns. Proc. Assoc. Off. Seed Anal. 64:109-114.
13. _____. 1976. Effects of gibberellin on germination of forest tree seeds with shallow dormancy. In IUFRO Int. Symp. on Physiology of Seed Germination. Fuji, Japan. p. 21-32.
14. _____. 1976. Maturation of Shumard and white oak acorns. For. Sci. 22:149-154.
15. _____. 1980. Collection, conditioning and certification of forest tree seed. In Advances in Forest Genetics, P. K. Khosla (ed.). Ambika Publ., New Delhi. ρ. 60-78.
16. _____. 1983. Germination response of loblolly pine to temperature differentials on a two-way thermo-gradient plate. J. Seed Tech. (In press).
17. _____ and T. R. Dell. 1976. The Weibull function: a new method of comparing seed vigor. J. Seed Tech. 1:96-103.

18. Bonner, F. T. and R. E. Farmer, Jr. 1966. Germination of sweetgum in response to temperature, moisture stress, and length of stratification. For. Sci. 12:40-43.

19. _____ and J. A. Vozzo. 1983. Measuring southern pine seed quality with a conductivity meter--does it work? Proc. Southeast. Forest Nurserym. Conf. (In press).

20. Bould, A. and B. K. Abrol. 1981. A model for seed germination curves. Seed Sci. & Tech. 9:601-611.

21. Campbell, R. K. and F. C. Sorenson. 1979. A new basis for characterizing germination. J. Seed Tech. 4:24-34.

22. Ching, T. M. 1973. Adenosine triphosphate content and seed vigor. Pl. Physiol. 51:400-402.

23. Clatterbuck, W. K. and F. T. Bonner. 1983. Utilization of food reserves in Quercus seed during storage. Seed Sci. & Tech. (In press).

24. Czabator, F. J. 1962. Germination value: an index combining speed and completeness of pine seed germination. For. Sci. 8:386-396.

25. Donald, D. G. M. 1981. Dormancy control in Pinus patula seed. South Afr. For. J. 118:14-19.

26. Dunlap, J. R. and J. P. Barnett. 1982. Germination characteristics of southern pine as influenced by temperature. In Proc. South. Cont. For. Tree Seedl. Conf. USDA For. Serv. Gen. Tech. Rpt. SO-37. p. 33-36.

27. Edwards, D. G. W. 1973. Effects of stratification on western hemlock germination. Can. J. For. Res. 3:522-527.

28. _____. 1978. Maturity and seed quality. In Proc. Flowering and Seed Development in Trees: A Symposium. IUFRO, Starkville, USA. p. 272-280.

29. Ellis, R. J. and E. H. Roberts. 1980. Improved equations for the prediction of seed longevity. Ann. Bot. 45:13-30.

30. Fechner, G. J., K. E. Burr, and J. F. Myers. 1981. Effects of storage, temperature, and moisture stress on seed germination and early seedling development of trembling aspen. Can. J. For. Res. 11:718-722.

31. Franklin, J. F. 1974. Abies Mill., Fir. In Seeds of Woody Plants in the United States. USDA For. Serv. Ag. Handb. 450. p. 168-183.

32. Goodchild, N. A. and M. G. Walker. 1971. A method of measuring seed germination in physiological studies. Ann. Bot. 35:615-621.

33. Grabe, D. R. 1964. Glutamic acid decarboxylase activity as a measure of seedling vigor. Proc. Assoc. Off. Seed Anal. 54:100-109.

34. Haavisto, V. F. 1980. The retention of seed viability in semi-serotinous black spruce cones. In IUFRO Int. Symp. on Forest Tree Seed Storage. Petawawa, Canada. p. 91-95.

35. Haddon, B. D. and D. A. Winston. 1980. Germination after two years storage of artificially ripened white spruce seed. In IUFRO Int. Symp. on Forest Tree Seed Storage. Petawawa, Canada. p. 75-80.

36. Hare, R. C. 1981. Nitric acid promotes pine seed germination. USDA For. Serv. Res. Note SO-281. 2 p.

37. International Seed Testing Association. 1976. International rules for seed testing 1976. Seed Sci. & Tech. 4:1-177.

38. _____. 1981. Handbook of vigour test methods. Int. Seed Test. Assoc. Zurich. 72 p.

39. _____. 1983. Report of the vigour committee. 20th Int. Seed Test. Assoc. Congress. Ottawa, Canada. 12 p.

40. Janssen, J. G. M. 1973. A method of recording germination curves. Ann. Bot. 37:705-708.

41. Jones, L. 1962. Ninth annual report. Eastern Tree Seed Laboratory. USDA For. Serv., Southeast. For. Expt. Stn. 18 p.

42. Khan, A. A. (ed.). 1977. The physiology and biochemistry of seed dormancy and germination. Elsevier Biomedical Press. 447 p.

43. King, M. W. and E. H. Roberts. 1979. The storage of recalcitrant seeds--achievements and possible approaches. Int. Board for Plant Genet. Res. Rome. 96 p.

44. Leadem, C. 1980. Seed viability of Abies, Picea, and Tsuga after storage in the cones. In IUFRO Int. Symp. on Forest Tree Seed Storage. Petawawa, Canada. p. 57-67.

45. Mayer, A. M. and A. Poljakoff-Mayber. 1975. The germination of seeds. Pergamon Press Ltd., Oxford. 2nd Ed. 192 p.

46. McLemore, B. F. 1969. Long stratification hastens germination of loblolly pine seed at low temperatures. J. For. 67:419-420.

47. _____ and F. J. Czabator. 1961. Length of stratification and germination of loblolly pine seed. J. For. 59:267-269.

48. Miller, T. and D. L. Bramlett. 1978. Damage to reproductive structures of slash pine by two seed-borne pathogens: Diplodia gossypina and Fusarium moniliforme var. subglutinans. In Proc. Flowering and Seed Development in Trees: A Symposium. IUFRO. Starkville, USA. p. 347-355.

49. Moore, R. P. 1969. History supporting tetrazolium seed testing. Proc. Int. Seed Test. Assoc. 34:233-242.

50. Moore, M. B. and F. A. Kidd. 1982. Seed source variation in induced moisture stress germination of ponderosa pine. USDA For. Serv. Tree Plant. Notes 33(1). p. 12-14.

51. Moore, F. D. and E. E. Roos. 1982. Determining differences in viability loss rates during seed storage. Seed Sci. & Tech. 10:283-300.

52. Muller-Olsen, C. and M. Simak. 1954. X-ray photography employed in germination analysis of Scots pine (Pinus silvestris L.). Medd. F. Strat. Skogsforsk. Instt. 44(6). p. 1-19.

53. Pitel, J. A. 1980. Accelerated aging studies of seeds of jack pine (Pinus banksiana Lamb.) and red oak (Quercus rubra L.). In IUFRO Int. Symp. on Forest Tree Seed Storage. Petawawa, Canada. p. 40-54.

58

54. Rediske, J. H. and D. C. Nicholson. 1965. Maturation of noble fir seed--a biochemical study. Weyerhaeuser Forestry Pap. No. 2. Weyerhaeuser Co., Tacoma, Wash. 15 p.

55. Roberts, E. H. 1972. Storage environment and the control of viability. In Viability of Seeds. (E. H. Roberts, ed.). Syracuse Univ. Press. p. 14-58.

56. Roos, E. E. 1982. Induced genetic changes in seed germplasm during storage. In The Physiology and Biochemistry of Seed Development, Dormancy and Germination. (A. A. Khan, ed.). Elsevier Biomedical Press. p. 409-434.

57. Ryynanen, M. M. 1982. Individual variation in seed maturation in marginal populations of Scots pine. Sil. Fenn. 16:185-187.

58. Schimpf, D. J., S. D. Flint, and I. G. Palmblad. 1977. Representation of germination curves with the logistic function. Ann. Bot. 41:1357-1360.

59. Simak, M. 1966. [Chromosome changes in aging seed.] Skogen. 53:28-30. [Swed.] [Abstr. in For. Abs. 27:3766. 1966.]

60. _____. 1976. Germination improvement of Scots pine seeds from circumpolar regions using polyethylene glycol. In IUFRO Int. Symp. on Physiology of Seed Germination. Fuji, Japan. p. 145-153.

61. _____. 1981. [Removal of filled-dead seeds from a seed bulk.] Sver. Skogsvard. Tidsk. 5:31-36. [Swed.].

62. Stein, W. I. 1965. A field test of Douglas-fir, ponderosa pine, and sugar pine seeds treated with hydrogen peroxide. USDA For. Serv. Tree Plant. Notes. 71:25-29.

63. Suszka, B. T. Tylkowski. 1980. Storage of acorns of the English oak (Quercus robur L.) over 1-5 winters. Arbor. Korn. 25:199-229.

64. Tanaka, Y. 1980. Effect of freezing temperatures on Noble fir seed during cone storage. In IUFRO Int. Symp. on Forest Tree Seed Storage. Petawawa, Canada. p. 68-74.

65. Villiers, T. A. 1974. Seed aging: chromosome stability and extended viability of seeds stored fully imbibed. Pl. Physiol. 53:875-878.

66. Vozzo, J. A. 1974. Special radiographic techniques. In Proc. X-ray Symp. USDA For. Serv., Macon, GA. p. 79-83.

67. _____. 1978. Carbohydrates, lipids, and proteins in ungerminated and germinated Quercus alba embryos. For. Sci. 24:486-493.

68. _____. 1983. Pericarp changes observed during Quercus nigra L. germination. Seed Sci. & Technol. (In press).

69. _____ and R. W. Young. 1975. Carbohydrate, lipid, and protein distribution in dormant, stratified, and germinated Quercus nigra embryos. Bot. Gaz. 136:306-311.

70. Wasser, R. G. 1979. Tree improvement research. In Proc. 1978 Southeast. Forest Nurserymen's Conf. USDA For. Serv. Tech. Publ. 5A-TP6. p. 15-25.

71. Webb, D. P. and V. P. Agnihotri. 1970. Presence of a fungal inhibitor in the pericarps of Acer saccharum fruits. Can. J. Bot. 48:2109-2116.
72. Wilcox, J. R. 1968. Sweetgum seed stratification requirements related to winter climate at seed source. For. Sci. 14:16-19.
73. Woodstock, L. W. and D. F. Grabe. 1967. Relationships between seed respiration during imbibition and subsequent seedling growth in Zea mays L. Pl. Physiol. 42:1071-1076.
74. Wynens, J. and T. Brooks. 1979. Seed collection from Loblolly pine. Ga. For. Res. Pap. 3. n.p.

4. MANIPULATING LOBLOLLY PINE (PINUS TAEDA L.) SEED GERMINATION WITH SIMULATED MOISTURE AND TEMPERATURE STRESS

J. R. DUNLAP* AND J. P. BARNETT
Weyerhaeuser Company, Hot Springs, Arkansas 71901 and Principal Silviculturist USDA-FS Alexandria, Louisiana 71360

ABSTRACT

Temperature and available moisture are environmental variables with substantial control over seed germination. Speed and completeness of germination in loblolly pine (Pinus taeda) seed strongly influence seedling yields. The impact of moisture and temperature on loblolly seed performance was investigated by germinating seed at temperatures cycling between optimal ($21^{\circ}C$) and stressful ($13^{\circ}C$ or $35^{\circ}C$) or in the presence of osmotically simulated moisture stress ranging between -300 kPa and -1500 kPa. The rate of germination was significantly reduced by exposure to the low temperature treatment. Any level of osmotically induced moisture stress decreased the rate and, ultimately, total germination. The impact of temperature or moisture stress was partially mitigated by increasing the stratification interval prior to germination. Consequently, low temperatures and/or decreased availability of moisture represent potential environmental restrictions on crop establishment and subsequent seedling yields. Germination patterns of seed experiencing environmental stress can be improved by extending the interval of stratification prior to sowing.

4.1 INTRODUCTION

During the next 20 years, most companies in the forest industry will rely on seed orchards as the source of all

---

*Currently USDA-ARS, Research Scientist, Post Office Box 267, Weslaco, Texas 78596

Duryea, M.L. and Brown, G.N. (eds.). Seedling physiology and reforestation succes
©1984, Martinus Nijhoff/Dr W. Junk Publishers, Dordrecht/Boston/London. ISBN 978-94-009-6139-5

regeneration material. Tree improvement programs, particularly in the southeast, are demonstrating substantial gains in yield through selective breeding programs directed primarily at loblolly pine (Pinus taeda L.) (1). Seed production from orchards is limited and can fluctuate appreciably over time (34). Companies that make annual seed selections according to family performance will continually face limitations on available seed. Companies with regeneration demands in excess of annual orchard seed production capacities will encounter similar limitations. In essence, implementation of the tree improvement concept via production orchards will ultimately create more frequent situations where seed is limited. Consequently, the economic success of a company's tree improvement program will be strongly influenced by its ability to obtain a maximum number of progeny from limited supplies of seed.

The presence or absence of germination represents an easily measured indicator of seed quality with obvious effects on a seedling production system. However, seed quality and performance can differ substantially between seedlots with similar percentages of final germination. A more subtle measure of seed quality is rate of germination. Total germination and the rate at which it is achieved represent more accurate measures of seed quality. Consequently, Czabater (9) has developed a mathematical model to accurately describe seed germination which incorporates both total and rate of germination into a single equation. The equation results in a numerical rating called germination value (GV) with the larger numbers implying faster and more complete germination. Loblolly germination like many other plant species generally follows a sigmoidal pattern characterized by a lag period prior to slow initial germination followed by a rapid rise and finally plateauing out as germination for the entire population is completed. Considering the sigmoidal shape of the curve, the time required for 50% of a population to germinate represents a potentially reliable measure of relative seed quality in loblolly. For the purposes of this study, the time to 50% germination has been designated as G_{50} and merely represents a simple method for describing relative rates of germination.

The importance of relative differences in germination rates among seedlots is actually realized as a difference in rates of emergence in the nursery bed. Seedlots which germinate slowly have broad emergence patterns and result in similar variation of plant sizes within the germinant population. Early differences in plant size set up the potential for competitive advantages in the seedling population that can be sustained throughout the growing season (28,32). A heterogeneous population of phenotypes is difficult to manage and may result in a decrease of average seedling quality. Therefore, desirable germination patterns are characterized by a rapid and complete germination.

Rapid and complete germination is a desirable objective in theory however, a variety of external influences can readily distort even the best patterns. Storage and handling prior to sowing (4) and the germination environment (23) can play major roles in determining seed performance levels. In most outdoor growing systems, temperature and soil moisture are important in crop establishment (23,24). Interactions between temperature and seed germination have been documented for numerous plant species including corn (31), cocklebur (14), white ash (6), carrots (16), lettuce (11), soybeans (13), white spruce (15) and Norway spruce (33). In most cases the optimum temperature range varied with genotype and species. McLemore (26) and Barnett (2) have studied the response of loblolly seed to a range of constant temperatures. Total germination decreased below 18°C and above 30°C. Bonner (5) used a thermogradient plate to generate a temperature response profile for loblolly. Constant temperatures of 25°C to 28°C resulted in the most rapid, complete germination. Continuous exposure to temperatures below 24°C and above 32°C reduced germination rates by 45 to 75%. However, some of the higher germination rates were actually obtained by alternating between the temperature extremes. The temperature response appears to be a cumulative phenomenon with lethal thresholds more rapidly achieved under constant temperature regimes. Both Barnett (2) and Bonner (5) found that stratification improved the germination of seed subjected to stressful temperatures.

Germination patterns can also be distorted by exposing seed to different levels of available moisture. Minimum levels of tissue hydration must be achieved before biochemical and physical processes leading to germination can take place (20, 24). The moisture content of a seed is governed by moisture availability and capacity to take up available water. In certain hard-seeded species, moisture uptake is restricted by the seedcoat (19,22). However, full imbibition of loblolly seed can take place within 24 h and does not represent a barrier to germination (unpub. data). Consequently, moisture availability after sowing plays a critical role in controlling germination patterns of loblolly. Uhvites (35) used mannitol to simulate moisture stress in studies on alfalfa seed germination. He found that germination was reduced at water potentials less than -700 kPa. Similar studies have been conducted with calabrese (20), corn (30), chickpea (17), wheat (12), ponderosa pine (10), longleaf and slash pine (3). In every study, the response to decreasing water potential was a reduction in both rate and total germination of the sample population. Considering the available information, studies were conducted to better characterize the influences of temperature and moisture on the germination patterns in several half-sib loblolly orchard families. These studies also investigated the use of stratification as a method for manipulating germination patterns.

4.2 PROCEDURE

4.2.1 Material and methods

4.2.1.1 Sample preparation. Seed were selected from half-sib families located in Weyerhaeuser orchards at Eutaw, Alabama and Washington, North Carolina. Germination tests consisted of three replicates containing 50 seed each. Seed were germinated on one layer of Kimpak paper in 10cm x 10cm plastic boxes sealed

with fitted lids. Germination was checked daily or every other
day depending on the projected rate of germination. A seed was
considered germinated when the protruding radicle had initiated
geotropic curvature. Stratification was accomplished by chilling
fully imbibed seed in 60 ml plastic bags at 2°C for various time
intervals. All treatment comparisons were germinated
simultaneously.

Table 1. Temperature treatments expressed in daily degree-hours
an actual time-temperature relationships for each regime

Treatment Number	Temperature Regime[1] (daily deg-hrs)	Temperature (°C)		
		13	22	35
		-------hrs------		
1	420	12	12	0
2	438	10	14	0
3	456	8	16	0
4	538	0	24	0
5	632	0	16	8
6	658	0	14	10
7	684	0	12	12

[1] 12-h photoperiod centered within the warmest portion of
each cycle.

4.2.1.2 Temperature stress. Stratified and unstratified seed
were subjected to temperature extremes of 13°C or 35°C. Stress
levels were controlled by alternating with a temperature considered
to be optimum for germination (22°C). The exposure time to stressful temperatures was 8, 10, or 12 h (Table 1). Daily deg-hrs were
calculated for each day by multiplying the hours spent at each
temperature by the numerical value of that temperature and summing
over a 24-h cycle. This procedure provided a measure of the
cummulative exposure experienced at each temperature regime.

4.2.1.3 Moisture stress Moisture stress was developed with
polyethylene glycol (PEG) 6000 according to the procedure
described by Michel and Kaufman (29). Concentrations of PEG 6000
in distilled water were adjusted to provide osmotic potentials of

-300, -600, -1000 and -1500 kPa at 22°C. Seed were placed on Kimpak pads of equal weight and wetted with 35 ml of the appropriate PEG solution. Germination boxes were then sealed and placed in an incubation chamber at constant 22°C. All PEG solutions were freshly prepared for each experiment from the same batch of chemical.

4.2.1.4 Data analysis. Final germination was defined as the percentage of total population germinated within 30 days (d). Germination rates were used as a measure of seed vigor and defined as the number of days required for 50% of population to germinate (G_{50}). Portions of a day were determined by interpolation. Data points consist of the mean for three replicates plus or minus one standard error which has been incorporated into the boundary of each symbol.

4.3 RESULTS

4.3.1 Temperature stress

Seed that were incubated under suboptimal temperature regimes (treatments 1, 2 and 3) germinated with G_{50} values of 9 and 7 d for the southern and northern sources, respectively (Table 2). With increasing daily deg-hrs, the speed of germination began to increcase as indicated by decreasing values for G_{50}. The most rapid and uniform germination for both sources took place in treatments 5, 6 and 7 which included periodic exposure to the highest temperature, 35°C. Final germination was

Table 2. Germination rates (G_{50}) for a southern (Texas) and Northern (Arkansas) source of loblolly incubated under seven different temperature regimes

Treatment	Temperature Regime (daily deg-hrs)	Germination Rate (G_{50})	
		Southern Source	Northern Source
1	420	9 ± 0	7 ± 0.4
2	438	10 ± 0.2	8 ± 0.2
3	456	7.6 ± 0.4	6.5 ± 0
4	538	6.3 ± 0.3	5.3 ± 0.3
5	632	4.7 ± 0.2	3.5 ± 0
6	658	5.2 ± 0.2	3.8 ± 0.2
7	684	6.0 ± 0.3	3.5 ± 0

unaffected by any of the temperature regimes and exceeded 94% in all treatments (unpub. data).

The influence of stratification on seed germination was evaluated at all temperature regimes. The treatment differences due to stratification decreased with exposure to regimes incorporating higher temperatures. Consequently, the description of stratification response was restricted to the low temperature treatment (420 daily deg-hrs). Without stratification, neither source had achieved 50% germination after 30 d of incubation (Fig. 1). Seed stratified only 20 d reached 50% germination (G_{50}) in 9 and 11 d for the northern and southern sources, respectively. With continued stratification for up to 60 d, both sources responded with decreasing G_{50} values. In addition to more rapid germination, the difference in G_{50} values between sources was reduced from 2 to less than 1 day.

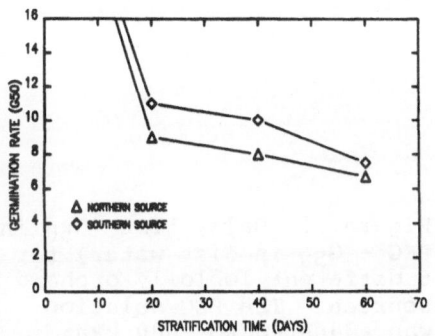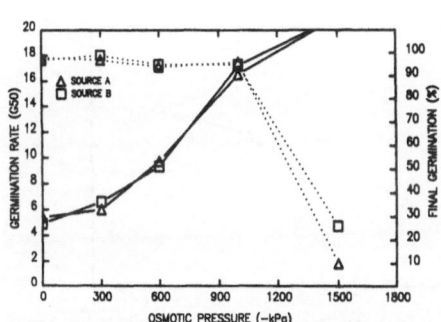

Figure 1. (left) Changes in the rate of germination (G_{50}) with increasing stratification time for an Arkansas (northern) and Texas (southern) orchard source of loblolly pine seed.
Figure 2. (right) The rate (-----) and final (...) germination of loblolly orchard sources incubated in PEG solutions of various water potentials.

4.3.2 Moisture stress

The rate of germination (G_{50}) was slowed as the osmotic potential of the incubation medium decreased from -300 kPa to

−1000 kPa (Fig. 2). The initial G_{50} determined in distilled water was approximately 5 d but slowed to 17 d at −1000 kPa. Major delays in G_{50} occurred at osmotic potentials exceeding −300 kPa. At −1500 kPa, the sample populations from both sources failed to reach 50% germination within 30 d. Final germination for up to and including −1000 kPa ranged between 95 and 97% (Fig 2). However, incubation at −1500 kPa decreased final germination after 30 d to 10 and 16% for source A and B, respectively.

Only two genotypes were used in the initial experiment and responded in a similar fashion to increasing levels of osmotic stress. Because of the possible influence of genotype on response, six different genotypes were screened simultaneously to determine the relative response to a common level of moisture stress. The difference in G_{50} determined in distilled water and at −600 kPa was calculated and referred to as delay time. Delay times for the six sources ranged from 3 to 6 d (Fig. 3).

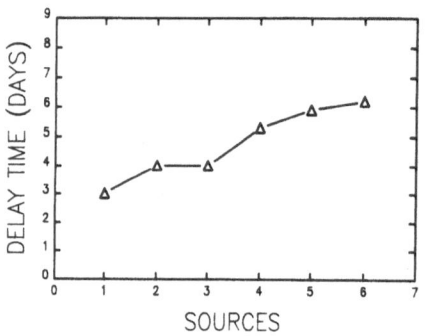

Figure 3. Delay time (G_{50} in PEG − G_{50} in dist water) for 6 different loblolly orchard sources. The PEG solution was adjusted to −600 kPa.

Stratification was subsequently investigated as a means of overcoming osmotically induced distortions of seed germination patterns. The rate of germination in distilled water improved with increased stratification time and reached a maximum rate after 60 to 80 d of stratification (Fig. 4). Similar increases in the rate of germination with stratification occurred in seed incubated at −600 kPa. However, the speed of germination at

-600 kPa was decreased substantially in both sources after 60 or
80 d of stratification. The interaction between moisture stress
and stratification was also determined with regard to final
germination at -1000 kPa, the approximate threshhold level for
significant inhibition (Fig. 5). The final germination of seed
was reduced significantly by stratification in excess of 40 d.
However, no differences among stratification treatments were
detected when incubation took place in distilled water.

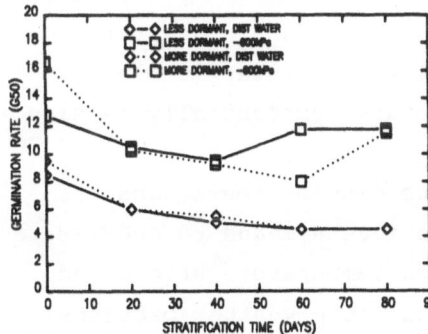 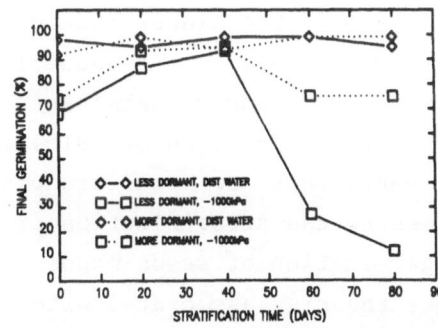

Figure 4. (left) The effect of increasing stratification time on
germination rates (G_{50}) of two loblolly orchard sources
incubated in the presence (PEG, -600 kPa) and absence (dist water)
of moisture stress.
Figure 5. (right) The effect of increasing stratification time on
final germination of two loblolly orchard sources incubated in the
presence (PEG, -1000 kPa) and absence (dist water) of moisture
stress.

4.4 DISCUSSION

The germination patterns of several loblolly orchard families
as described by G_{50} and final germination percentage were
altered by changes in temperature and osmotic potential. Although
final germination was unaffected, increasing the diurnal exposure
time to low temperatures resulted in slower germination (Table 2).
As seed accumulated daily exposure to warm temperatures, the rate

of germination was accelerated and produced a more uniform and
desirable pattern. The experimental temperature regimes simulated
conditions found in southeastern seedling nurseries where sowing
begins in early April. Therefore, the undesirable effects of
temperature on uniformity of emergence are primarily a function of
accummulated exposure to suboptimal temperatures (25). Curiously,
nursery managers most often refer to late sowing and high
temperatures when describing historical problems associated with
germination (pers. obs.). The data presented in this paper and
others (2,5) prove that lethal high temperature threshholds are
not achieved under field conditions normally encountered during
germination. The higher risk of crop losses should occur through
extended exposure to unseasonably cool temperatures (Table 2).
The resulting crop of germinants will vary substantially in size
due to erratic emergence. Slow emergence may also create
germinants that are more susceptable to disease (pers. obs.).
Increasing the stratification time provides a means to accelerate
the germination of seeds exposed to low temperature stress and
reduce the risk associated with unpredictable weather patterns
(Fig. 1).

Low water potentials had a relatively large influence over
loblolly seed germination. The principal impact was to slow
germination rates and broaden the emergence intervals at osmotic
potentials less than -300 kPa (Fig. 2). Final germination of
sample populations was unaffected until water potentials reached
-1000 kPa or greater. Many of the operational seedling nurseries
have been selected on the basis of soil texture with relatively
high sand contents being a desirable characteristic. However, the
high percentage of sand in these soils reduces their water holding
capacity which permits broad shifts in soil water potential. This
can be especially true in the upper 2 cm of an unmulched seedbed.
It is not unreasonable to speculate that soil water potentials
routinely approach or exceed -1000 kPa even with intensive
management. The magnitude of fluctations in soil water potential
can be augmented by high temperatures, low humidity and wind.

These conditions are common to spring weather patterns in the mid-south and southeast. Therefore, soil water potential rather than high temperatures may represent the major constraint to maximizing germination.

As previously described in the temperature experiments, stratification was also effective in overcoming the reduction in germination rates caused by low osmotic potentials (Fig. 4 and 5). However stratification in excess of 40 d decreased the rates and final germination of certain treatments. This result was totally unexpected and inconsistent with results from field tests and standard germination tests (unpub. data). The results may reflect a phenomenon similar to accelerated aging (18) or storage at high moisture contents (27) previously reported for seeds. Coolbear, et al. (8) observed a decrease in germination of tomato seeds with prolonged incubation at water potentials of -700 to -1000 kPa. They attributed this effect on germination to microbial degradation but also referred to losses of food reserves during incubation. Mobilization and metabolism of critical food reserves may explain losses of seed vigor associated with seed storage; extended metabolism of critical food reserves may explain losses of seed vigor associated with seed storage, extended stratification and prolonged exposure to water stress. Stratification apparently overcomes barriers to germination imposed by low osmotic potentials but only when growth of the radicle is not severely impaired. Under field conditions, seed may experience cyclic exposure to severe water potentials but not continuously as in our experiments. Therefore, seed will be exposed to intervals of higher water potentials permitting radicle growth and, ultimately, germination. The results do imply that minimization of any moisture stress encountered by seed during germination will improve the potential for uniformity of emergence.

The effects of temperature and moisture on germination can be applied qualitatively to loblolly as a species. However, the specific performance characteristics of individual genotypes can not be predicted (Fig. 3). Germination characteristics in response to specific environmental conditions should be determined for each individual family with consideration given to harvest year.

Maximizing yields from limited quantities of high value seed will require a greater understanding of how the seed interact with the environment. For example, membrane and enzyme degradation have been proposed as explanations for loss of germination potential in various seeded species when exposed to environmental stresses (21). Although temperature and moisture are difficult to control in large production systems, conditions can be optimized with precision technology and an understanding of biological responses. The real advances will come through manipulating the biological potential of a seed to germinate under adverse conditions. The basic information supporting such strategies is extremely limited. Carpita, et al. (7) recently obtained results providing insight into the possible origin of the stratification response in loblolly. The phenomenon of seed temperature/moisture response in loblolly can only be understood and subsequently manipulated through an extension of the current knowledge with emphasis on the physiological mechanisms controlling germination.

REFERENCES

1. Anon. 1981. Twenty-fifth Ann. Rept. N. C. State Univ.-Indust. Coop. Tree Improvement Program. 62 pp.
2. Barnett, J. P. 1979. Germination temperatures for container culture of southern pines. S. J. Appl. For. 3: 13-14.
3. Barnett, J. P. 1969. Moisture stress affects germination of longleaf and slash pine seeds. For. Sci. 15:275-276.
4. Barnett, J. P. and B. F. McLemore. 1970. Storing southern pine seeds. J. For. 68:24-27.
5. Bonner, F. T. 1983. Germination response of loblolly pine to temperature differentials on a two-way thermogradiant plate. Seed Sci. & Technol (In press).
6. Bonner, F. T. 1975. Germination temperatures and prechill treatments for white ash (Fraxinus americana L.), p. 60-65. In: Proc. Assoc. Off. Seed Anal., Vol. 65.
7. Carpita, N., J. Skaria, J. Barnett, and J. Dunlap. 1983. Influence of stratification on growth of radicles of loblolly pine (Pinus taeda L.) embryos. Physiol. Plant. (In press).

8. Coolbear, P., D. Grierson, and W. Heydecker. 1980. Osmotic
    pre-sowing treatments and nucleic acid accumulation in
    tomato seeds (Lycopersicon lycopersicum). Seed Sci. &
    Technol. 8:289-303.
9. Czabator, F. J. 1962. Germination value: an index
    combining speed and completeness of pine seed germination.
    For. Sci. 14:219-221.
10. Djavanshir, K. and C. P. Reid. 1975. Effect of moisture
    stress on germination and radicle development of Pinus
    elderica Medu. and Pinus ponderosa Laws. Can. J. For.
    Res. 5:80-83.
11. Dunlap, J. R. and P. W. Morgan. 1977. Reversal of induced
    dormancy in lettuce by ethylene, kinetin and gibberellic
    acid. Plant Physiol. 60:222-224.
12. El-Sharkawi, H. M. and I. Springuel. 1977. Germination
    of some crop plant seeds under reduced water potential.
    Seed Sci. & Technol. 5:677-688.
13. Emerson, B. N. and H. C. Minor. 1979. Response of
    soybeans to high temperature during germination. Crop
    Sci. 19:553-556.
14. Esashi, Y., Y. Taksukada, and Y. Ohhara. 1978.
    Interrelation between low temperature and anaerobiosis in
    the induction of germination of cocklebur seed. Aust. J.
    Plant Phys. 5:337-345.
15. Fraser, J. W. 1971. Cardinal temperatures for
    germination of six provenances of white spruce seeds.
    Dept. of Fisheries and Forestry, Can. For. Ser. Publ. No.
    1290, 10 pp.
16. Gray, D. 1979. The germination response temperature of
    carrot seeds from different umbels and times of harvest of
    the seed crop. Seed Sci. & Technol. 7:169-178.
17. Hadas, A. 1977. A suggested method for testing seed
    vigour under water stress in simulated arid conditions.
    Seed Sci. & Technol. 5:519-525.
18. Halloin, J. M. 1975. Solute loss from deteriorated
    cottonseed: relationship between deterioration, seed
    moisture and solute loss. Crop Sci. 15:11-15.
19. Hartmann, H. T. and D. E. Kester. 1975. Principles of
    propagation by seed, p. 108-145. In: Plant Propagation
    - Principles and Practices. Prentice Hall, New Jersey.
20. Hegarty, T. W. and H. A. Ross. 1978. Some characteristics
    of the water-sensitive process in the inhibition of
    germination by water-stress. Ann. Bot. 72:1222-1226.
21. Heydecker, W. 1977. Stress and seed germination: an
    agronomic view, p. 253-259. In: A. A. Kahn (ed.). The
    Physiology and Biochemistry of Seed Dormancy and
    Germination. Elsevier/North Holland Biomedical Press,
    N. Y.
22. Kahn, A. A. 1977. Seed dormancy: changing concepts and
    theories, p. 29-50. In: A. A. Kahn (ed.). The Physiology
    and Biochemistry of Seed Dormancy and Germination.
    Elsevier/North Holland Biomedical Press, N. Y.

23. Koller, D. 1972. Environmental control of seed germination, p. 2-102. In: T. T. Kozlowski (ed.). Seed Biology, Vol II. Academic Press, N. Y.

24. McDonough, W. T. 1977. Seed physiology, VI., p. 155-184. In: R. E. Sosebee (ed.). Rangeland and Plant Physiology., Soc. Range Mgt.

25. McLemore, B. F. 1969. Long stratification hastens germination of loblolly pine seed at low temperatures. J. For. 67:419-420.

26. McLemore, B. F. 1966. Temperature effects on dormancy and germination of loblolly pine seed. For. Sci. 12:284-289.

27. McLemore, B. F. and J. P. Barnett. 1968. Moisture content influences dormancy of stored loblolly pine seed. For. Sci. 14:219-220.

28. Mexal, J. G. 1980. Growth of loblolly pine seedlings. I. Morphologica variability related to day of emergence. Weyer. For. Res. Tech. Rept., No. 042-2088/80/40. pp. 10.

29. Michel, B. E. and M. R. Kaufmann. 1973. The osmotic potential of polyethylene glycol 6000. Plant Physiol. 51: 914-916.

30. Parmer, M. T. and R. P. Moore. 1968. Carbowax 6000, mannitol, and sodium chloride for simulating drought conditions in germination studies of corn (Zea mays L.) of strong and weak vigor. Agron. J. 60:192-195.

31. Riley, G. 1981. Effects of high temperature on the germination of maize (Zea mays L.). Planta 151:68-74.

32. Ross, M. A. and J. L. Harper. 1972. Occupation of biological space during seedling establishment. J. Ecol. 60:77-88.

33. Simak, M. and S. K. Kamra. 1970. Germination studies on Norway spruce (Picea abies) seed of different provenances under alternating and constant temperatures, p. 383-391. In: Proc. Int. Seed Test. Assoc., Vol. 35.

34. Talbert, J. T. 1981. One generation of loblolly pine tree improvement: results and challenges, p. 106-120. Proc. of 18th Can. Tree Improvement Assoc. Mtg. Duncan, B. C.

35. Uhvits, R. 1946. Effects of osmotic pressure on water absorption and germination of alfalfa seeds. Am. J. Bot. 33:278-285.

Stock Quality

Stock Types

- **Bareroot**—Papers 5 and 6
- **Container**—Papers 7 and 8

5. ALTERING SEEDLING PHYSIOLOGY TO IMPROVE REFORESTATION SUCCESS

M. L. DURYEA and K. M. McCLAIN

Assistant Professor and Leader, Nursery Technology
Cooperative, Department of Forest Science, Oregon State
University, Corvallis, Oregon 97331, U.S.A. and Research
Scientist, Ontario Tree Improvement and Forest Biomass
Institute, Northern Forest Research Unit, P.O. Box 2960,
Thunder Bay, Ontario, Canada P7B G5G.

ABSTRACT

Reforestation success depends largely upon the physiological and morphological preparedness of seedlings to survive and grow after planting. Frost hardiness, mineral
nutrition, and carbohydrate reserves are critical elements
of seedling physiology. This paper reviews these elements
to illustrate how they are affected by nursery cultural
practices and, in turn, their impact on field survival and
growth.

Inappropriate application or timing of some nursery practices, such as irrigation and fertilization, can lengthen
the growing season and delay the onset of dormancy; for some
species, such as Douglas-fir (Pseudotsuga menziesii [Mirb.]
Franco), extended summer growth can lead to fall frost
damage in the nursery. However, well-planned schedules of
irrigation and fertilization, as they relate to species phenology, can increase frost hardiness. Seedling damage that
does occur may or may not be visible in the nursery, and the
performance of the damaged seedlings after planting may be
severely reduced under some field conditions. Field performance also will be reduced in seedlings that are not fall-
hardy when lifted, processed and stored for planting, and in
seedlings that, for any number of reasons related to
cultural practices, lack resistance to subfreezing temperatures in the field.

Mineral nutrition of seedlings can be dramatically

Duryea, M.L. and Brown, G.N. (eds.). Seedling physiology and reforestation succes
©1984, Martinus Nijhoff/Dr W. Junk Publishers, Dordrecht/Boston/London. ISBN 978-94-009-6139-5

increased by growing irrigated seedlings at low seedbed densities and moderately high fertilizer levels. Nitrogen (N) concentrations in the range of 1.7 to 2.3% generally result in seedlings that exhibit enhanced field survival and height growth. The benefits of nutrition, however, can be altered by its effect on seedling size, degree of frost hardiness, and levels of carbohydrates.

Cold storage, an accepted and sometimes necessary nursery practice for holding seedlings between lifting and planting, often depletes carbohydrate reserves. Poor seedling survival after cold storage has, for a number of species, been attributed to depleted carbohydrate reserves. Because seedling root and shoot growth after field planting may also be dependent on carbohydrate reserves, depleted levels could have a marked impact on first-year performance in the field.

5.1 INTRODUCTION

Successful reforestation is highly dependent on the planting of high quality forest tree seedlings. Until recently, seedling quality has been defined mainly by the morphological characteristics or physical appearance of the seedling. However, field performance is also related to the seedling's physiological condition and interaction with the environment of the planting site. For this reason, there has recently been an increasing trend toward physiological characterization of planting stock (for example, 16,62,78); poor performance after planting has been linked to physiologically-dependent responses, such as inability to regenerate new roots, poor nutrient status or impaired water uptake. Silviculturists have recognized the need for a greater understanding of those aspects of seedling physiology that are critical to optimal field performance.

This paper reviews three major elements of seedling physiology that are known to have a significant impact on reforestation success: frost hardiness, mineral nutrition,

and carbohydrate reserves. For each element we review (1)
its principles and importance to seedling quality, (2) how
it may be altered by nursery practices, and (3) how the
altered physiological state affects field survival and
growth.

5.2 FROST HARDINESS

Frost hardiness can be defined as the minimum temperature
to which a tree seedling can be exposed without sustaining
damage. We should note that the terms "frost hardiness" and
"cold hardiness" are often considered equivalent and are
used interchangeably; however, we consistently use frost
hardiness in this paper. Also, the term "field performance"
is used to refer to survival and growth in the field.

A number of comprehensive reports review the stages of
frost hardiness acquisition, the biochemical and physiological changes associated with this acclimation, and techniques
for screening for frost hardiness (1, 13, 17, 29, 83, 95).
Most agree with Weiser (95), who states that woody plants
undergo three stages to become frost hardy, also called
acclimation or hardening. The first stage is triggered by
decreasing day length and usually accompanies the cessation
of growth in the fall. The second stage is induced by
freezing or near-freezing temperatures and involves changes
in plant compounds (such as sugars, proteins, amino acids,
nucleic acids) and results in the development of a
resistance to freezing. The third stage, triggered by even
lower temperatures, involves the binding of water and
enables the plant to resist dehydration (37, 95).

Tree seedlings reach maximum frost hardiness in midwinter. This maximum varies by tree species and by seed
sources within a species. Hardiness is lost when buds begin
to open in the spring; however, even unopened buds can be
injured should large sudden drops in temperature occur (37).
Figure 1 shows the hardening and dehardening patterns of
white pine (Pinus strobus L.), which has a maximum hardiness
level in midwinter below -40°C (29).

80

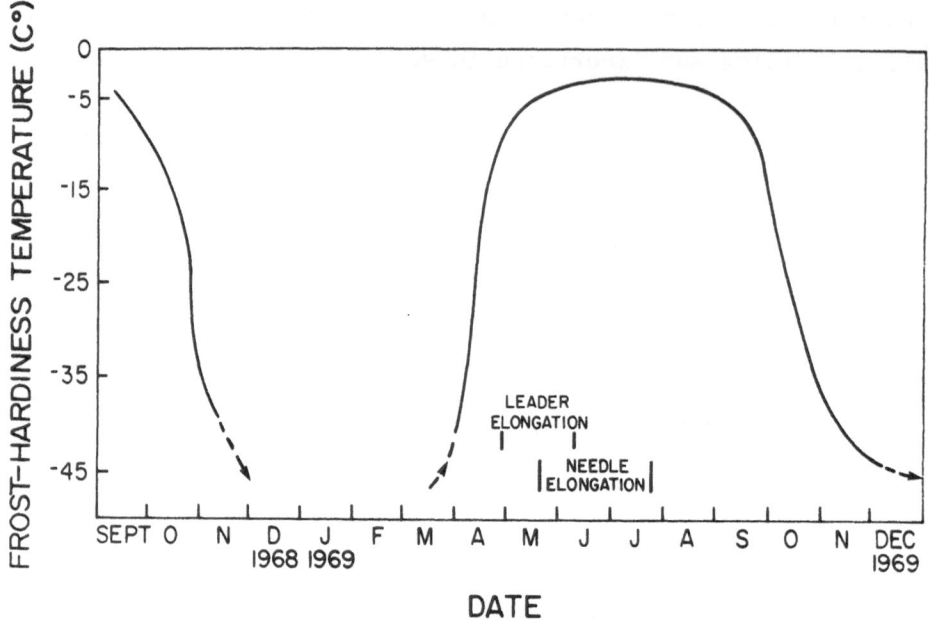

FIGURE 1. General pattern of frost hardiness of white pine
(<u>Pinus strobus</u> L.). Frost hardiness increased from
September to late November until it reached its maximum of
-40°C. Dehardening began in April. (Adapted from [29])

One question which has been continually asked is whether
seedlings must attain deep dormancy before they can develop
frost hardiness (37). The key factor, however, seems to be
growth cessation and not the development of dormancy,
because low temperatures can arrest growth and promote frost
hardiness without necessarily inducing dormancy (95). The
reverse can also occur; some southern pine species, for
example, become dormant but do not withstand low temperatures (37). However, most temperate zone species
acquire frost hardiness after they have entered the dormancy
sequence.

The frost hardiness of seedlings changes dramatically
during the year (95). Seedlings are most likely to be
damaged during several critical periods in a 2-year growing
regime in the nursery. Those critical periods are:

1. Spring (newly germinated seedlings)

2. Fall (especially when growth has been extended or
    lammas growth has occurred)

3. Spring (when trees resume growth during their
    second growing season)

4. Fall [(especially when growth has not ceased) of
    second growing season]

Frost damage during these periods depends on the low temperature the plant experiences, the rate of freezing, and
the duration of the low-temperature period. Damage can
range from slight needle damage, to top dieback, to seedling
mortality. Seedlings that are not killed will vary in their
ability to resume growth and attain quality standards
necessary to meet nursery yields. Often what results are
multiple-topped seedlings, which in many cases are culled.
Nursery managers are extremely concerned about cultural
practices that can reduce the physiological quality of
seedlings and in turn decrease nursery yield or field growth
and survival.

5.2.1 Cultural practices and frost hardiness

Extensive crop losses have convinced nurseries to change
many of their cultural practices in order to minimize frost
damage to seedlings. Weyerhaeuser Company, for example,
experienced a freeze in 1969 at its Mima nursery in
Washington that seriously damaged or killed 12 million 1+0
seedlings. Weyerhaeuser plotted the frequency of critical
freezing temperatures at the nursery and found that a severe
crop loss could be expected every 4 years (3). Many of the
practices discussed below have since been employed at many
nurseries, and have decreased seedling frost damage.

82

5.2.1.1 Irrigation regime. Irrigation in the spring and
early summer promotes growth of both new germinants and
recently flushed second- and third-year seedlings.
Continued irrigation throughout the summer promotes growth.
Seedlings will continue to grow and, if they do set a bud, a
second flush in the late summer or early fall is very
likely. This second flush can harm frost hardiness in two
major ways: 1) the new, recently grown plant tissue is non-
hardy and susceptible to fall frost damage (21) and 2) one
of the key factors in the induction of frost hardiness is
growth cessation; if this has not occurred, then acclimation
or frost hardening may not occur (95). For these reasons,
as well as to induce bud dormancy, many nurseries impose a
nursery irrigation schedule with a moderate stress regime at
the end of the desired growth period. Some of the results
are earlier budset (31), earlier induction of dormancy (42,
100), and increased frost hardiness of seedlings (12).
Blake et al. (12) demonstrated that a moderate stress regime
(-5 to -10 bars) improved frost hardiness of Douglas-fir
seedlings, whereas a higher stress level (-10 to -15 bars)
reduced frost hardiness to that of the control (Table 1).
Lammas growth occurred only in the control seedlings (0 to
-5 bars stress); the highest stress level (-10 to -15 bars),
although detrimental to frost hardiness, did induce budset.

TABLE 1. Mortality of nursery-grown Douglas-fir seedlings
subjected to a frost hardiness test. Moisture stress was
applied at three levels between late July and late August
(adapted from [12]).

Lifting date	Test temperature (°C)	Age class	Predawn stress level (bars)		
			0 to -5[a]	-5 to -10	-10 to -15
			-------Percent mortality-------		
October	-5.5	1+0	70.0	22.5	41.0
	-5.5	2+0	57.5	50.0	65.0
	-5.5	2+1	67.5	57.5	80.0
December	-14.8	1+0	40.0	5.0	50.0
	-29.3	2+0	20.0	5.0	35.0

[a]Control

The date of initiation of the moisture stress regime has also significantly affected frost hardiness. In the Blake et al. study (12), frost hardiness was greatest for seedlings that started receiving stress July 15. As the initiation date was delayed, frost hardiness decreased (Table 2). Other studies also show that the moderate stress regime should be imposed before the naturally inductive short days, and that a delay in this stress regime may inhibit frost hardiness (43, 80, 84).

Table 2. Effects of moisture stress initiation date on the frost hardiness of 2+0 Douglas-fir seedlings (adapted from [12]).

Test		Moisture stress initiation date				
Month	Temperature (°C)	July 15	Aug. 1	Aug. 15	Sept. 1	Control
		------------Percent mortality-----------				
October	-7	35	50	64[a]	68	69[b]
November	-21	49	50	67	73	---

[a]Adjacent means underscored by the same line do not differ significantly at the 5% level.

[b]Control means were excluded in the statistical analysis.

5.2.1.2 Fertilization. The application of N during the growing season can affect the development of frost hardiness by prolonging growth and delaying the onset of dormancy, both of which can result in frost damage to succulent shoots (82, 88, 97). However, other studies show that fertilization, especially in cases where the addition of N was not correlated with the continued growth of the seedlings, improves frost hardiness (e.g., late summer N application to black walnut ([Juglans nigra L.] seedlings [97] and fall N fertilization of Douglas-fir seedlings [81]). Timmis (82), in his work with Douglas-fir container seedlings, found that frost hardiness was more closely related to the ratio of potassium (K) and N than to the level of any individual element. Other authors have expressed the importance of K levels to frost hardiness (88, Larsen [1978] as cited in 88, 52).

One other possible effect on frost hardiness is suggested by Benzian et al. (11), who reported that a fall application of N to Sitka spruce (Picea sitchensis [Bong.] Carr.) seedlings caused earlier bud break the following spring and resulted in increased frost damage. Other species in this study (Norway spruce [Picea abies L. Karst], western hemlock (Tsuga heterophylla (Raf.) Sarg.], grand fir [Abies grandis (Dougl. ex D. Don) Lindl.], and lodgepole pine [Pinus contorta Dougl. ex Loud.]) exhibited no increase in injury following N fertilization.

5.2.1.3 Photoperiod. Shortening day length, although impractical for bareroot nurseries, is important in the natural frost-hardening process and has been tested with container seedlings. Norway spruce seedlings that experienced a shortened day length at the end of their growth period had increased frost hardiness and were more storable (69). Short photoperiods that allowed sufficient light for photosynthesis also aided the frost hardening of Douglas-fir seedlings (85).

5.2.1.4 Species, seed source, and nursery location. Species and seed sources within a species from northern latitudes or higher elevations often set bud earlier and are less susceptible to injury from early frosts (37). For example, black walnut seedlings from Kentucky, Tennessee and Alabama seed sources suffered four times as much damage as did seedlings from a Michigan source after all experienced a frost in an Indiana nursery (97). Douglas-fir seedlings from southern Oregon sources and growing in Corvallis, Oregon set bud later and experienced more frost damage than did seedlings from sources in Washington and northern Oregon (Figure 2) (18). An early fall frost at the USDA Forest Service Wind River Nursery in Washington damaged high elevation Douglas-fir seedlings the least and low elevation Cascade and Coastal range (Oregon) Douglas-fir seedlings the most (24). Seed source differences such as these often show up in a nursery after an unexpected frost.

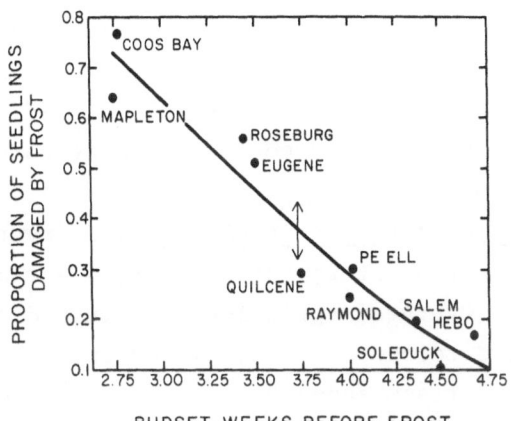

FIGURE 2. Proportion of Douglas-fir (<u>Pseudotsuga menziesii</u> [Mirb.] Franco) seedlings (from 10 seed sources) damaged by frost as related to the mean date of budset. (Vertical line is the 95% confidence limit at the overall mean budset.) (Adapted from [18])

A little-investigated frost hardiness problem is the effect of nursery location on frost hardening. A study in New Zealand compared the fall, winter, and spring frost hardiness of radiata pine (<u>Pinus radiata</u> D. Don) seedlings grown at seven nurseries. Seedlings from some of the nurseries consistently were less frost hardy than were seedlings from other nurseries. The two nurseries with the greatest differences in location produced frost hardiness measurements that were 5°C apart for the fall, 3°C for winter, and 4°C for spring (52). The influence of nursery location on the development of frost hardiness deserves more investigation, especially in areas such as New Zealand, where field mortality due to frost is common (92).

5.2.1.5 <u>Assessing frost hardiness</u>. Many nurseries assess frost hardiness of their seedlings and use the assessment for two purposes: 1) as a guide for frost protection during the fall or spring and 2) as an indicator of stock hardiness when seedlings are lifted and outplanted (62). Two steps are involved in assessing frost hardiness: first, seedlings are exposed to subfreezing temperatures, and second, seedling damage is evaluated (62, 83). The minimum temperature at which 50% of seedlings are killed (LT_{50}) is most often used for comparisons of hardiness. Timmis (83), Warrington and Rook (92), and Ritchie (62) provide the most recent reviews of techniques used to determine hardiness of forest planting stock.

5.2.1.6 <u>Irrigation for frost protection</u>. Many nurseries, especially in the northwestern United States, irrigate for frost protection. Overhead sprinklers are turned on when the air temperature reaches 0°C (32°F) and are left on until the air temperature rises above 0°C (from OSU Nursery Survey, described in [22]). The irrigation technique protects both newly germinated seedlings and older seedlings that are susceptible to spring and fall frost (3, 22).

5.2.1.7 <u>Other practices</u>. Other practices that promote earlier budset may also cause earlier induction of frost hardiness in the fall. These practices include earlier sowing of seed (76) and root wrenching (21). In contrast, some cultural practices induce lammas growth in the fall and increase the chance of frost damage; these practices include accidental lowering of seedbed density, which gives more nutrients and water to some seedlings in the nursery than to others, and late-season top pruning.

5.2.2 <u>Frost hardiness and field performance</u>

There are three times when frost hardiness of seedlings can be important to survival and growth: 1) when seedlings that have been frost damaged in the nursery are planted in the field, 2) when seedlings that are not frost hardy are lifted, 3) when seedlings that burst bud early in the spring after being planted in the field are then exposed to freezing temperatures. These three instances are discussed below.

5.2.2.1 <u>Damaged seedlings are planted in the field</u>. Characteristics of seedlings that have experienced frost

damage are browning of leaves, bud-kill, top-kill, or
complete mortality. Damaged seedlings left in the seedbeds
for another growing season often develop multiple tops, due
to bud and shoot damage. Although seedlings with multiple
tops are usually culled, little is known about their per-
sistence and growth in the field. One theory is that the
culling of visibly damaged (i.e., multiple-topped) seedlings
may select against seedlings with the highest growth potential (i.e., seedlings that flush twice during the year and
thus are frost damaged in the nursery) (24).

Seedlings may also be damaged in the nursery during the
fall before lifting. This damage--for example, bud-kill or
cambial damage--can escape detection when the seedlings are
graded.

The few studies that have investigated the field performance of frost-damaged seedlings have found harmful effects
on survival and growth. Edgren (24) reported that 51% of
frost-damaged Douglas-fir 2+0 seedlings failed to burst bud
the spring after field planting. The seedlings that failed
to burst bud grew less during the next 3 years in the field
than did seedlings that burst bud. In another study, 61% of
frost-damaged black walnut seedlings died in the first
growing season (97). Prevention of frost damage in the nursery seems to be of paramount importance to reforestation
success.

5.2.2.2 Seedlings are not frost hardy when lifted.
Seedlings that are not frost hardy when lifted may be
damaged by exposure to cold or subfreezing temperatures,
either during storage or in the field (33). Cold storage of
Douglas-fir seedlings too early in the fall is known to
jeopardize subsequent field survival and growth. van den
Driessche (90) notes that this may be partly explained by
inadequate development of frost hardiness that renders
seedlings less tolerant of storage at temperatures lower
than 2°C.

Freezer storage (at -1 to -2°C) is also becoming a more
commonly used practice; to ensure its success, tree

seedlings must be frost hardy when stored. Accidental freezing during cold storage (at 1°C) or shipping may also damage trees, with non-frost-hardy trees being especially susceptible. Non-hardy seedlings that are field planted and exposed to subfreezing field temperatures below their hardiness limit can also be damaged (62).

Frost injury incurred either during storage or after planting can range from slight needle damage to mortality. Such injury will in most cases inhibit growth; it may also weaken trees, making them more susceptible to other environmental stresses, such as drought and heat (92).

5.2.2.3 <u>Seedlings burst bud earlier the spring after planting</u>. Some cultural practices, such as fertilization, can change the phenology of seedlings. As previously mentioned, Sitka spruce that was fertilized with N in the nursery and then was field planted burst bud earlier in the spring and was frost damaged (11). Again, damage from early spring frosts could range from needle injury to seedling mortality.

5.3 MINERAL NUTRITION

North American nurseries used to practice the philosophy that seedlings should be produced under stressful conditions similar to those expected on the planting site. Nursery managers believed, for instance, that seedlings denied fertilizer would be accustomed to and better able to survive the low nutrient levels typical of many planting sites (96). This view, however, was dispelled when it became obvious that outplanted seedlings did not perform well unless nutrients, in either an organic or inorganic form, were added periodically to nursery beds (6).

Starvation nursery practices soon gave way to conscientious attempts to produce stock that was well-balanced, possessed good color, and was free of obvious defects. These attempts usually produced large seedlings but did not always result in successful plantations (72). Initial planting failures, however, were followed by concentrated

efforts to improve planting technique (for example, placement of roots) (66). Although not alone in his appreciation
of the problem, Wakeley (91) attributed the apparent incon-
sistencies in morphological characteristics of seedlings to
plantation performance. Wakeley concluded that a big
seedling was not necessarily a better seedling, and that
there must instead be a physiological grade that truly
represented a seedling's capability for survival.

During the last several decades, nutrient status of
seedlings has been recognized as a prominent component of
the physiological grade of seedlings at the time of
planting. This increasing recognition is not surprising,
considering the many physiological processes that require
the presence of the 16 macronutrients and micronutrients
essential to growth. Several excellent reviews (for
example, 19, 51, 53, 68) detail the specific roles of these
elements in physiological processes that regulate growth.
From these reviews, we infer that seedling growth can be
curtailed in the nursery and in the field if the plant does
not receive nutrients in sufficient amounts and in proper
proportions.

It follows, therefore, that forest nursery seedlings
whose nutritional status is deficient or imbalanced due to
inappropriate nursery management will not grow as well after
they are planted in the field. Nursery managers, therefore,
must consider the effects of commonly employed practices on
plant nutrition in order to ensure that nutrition does not
limit potential field performance.

This section reviews fertilization, seedbed density,
irrigation, and root wrenching because these practices can
effectively alter the nutritional status and growth of
seedlings, thereby affecting field performance. It should
be recognized at the outset that one practice may modify the
effect of another, and that it may be difficult to identify
the effect of a single, given practice. Moreover, nursery
practices in many cases markedly alter seedling morphology,
thereby necessitating consideration of both morphology and

nutrient status of planting stock in relation to field performance.

5.3.1 Cultural practices and plant nutrition

5.3.1.1 Fertilization. Application of organic or inorganic fertilizers is the most direct procedure to alter the nutritional status of seedlings. However, the motive for fertilization usually is the need to achieve a given seedling size, rather than to attain a particular nutrient concentration. Generally, if larger trees are required, the rate of growth of trees can be increased, within limits, by increasing fertilization (6).

As fertilization levels are increased, nutrient concentrations within the plant generally increase to an upper limit, providing other factors are not limiting. van den Driessche (87) assessed the effect of N and phosphorus (P) fertilizer treatments on the nutrient concentrations of 1+0 and 2+0 Douglas-fir; he found that as N (applied as ammonium nitrate) was increased from 0 to 100 kg/ha, 1+0 shoot N concentrations increased steadily from 1.09% to 2.01%, whereas P decreased from 0.17% to 0.12%. In 2+0 needles, N concentrations increased dramatically, from 0.71% (no N supply) to 2.75% at 200 kg/ha. P and K concentrations decreased with increasing N supply from 0.17% to 0.09% and 0.56% to 0.26% respectively. Increases in N concentrations also resulted in morphological changes. For example, shoot dry weight of 2-year-old seedlings increased from 8.8 g to 16.5 g as the application of N increased from 0 to 100 kg/ha; however, further increases in N produced only marginal weight gain.

Nutrient concentrations have also been examined for white spruce (Picea glauca [Moench] Voss) (5, 9, 49); without exception, the concentration of N in the foliage increased as the N fertilization level increased (Figure 3). There is, however, one reported exception. Armson (5) evaluated red pine (Pinus resinosa Ait.) in relation to fertilization levels of N, P and K; seedling dry weight increased dramatically from 2 to 6 g, but foliar N concentrations did not increase appreciably (Figure 3). This phenomenon was attributed to efficient N adsorption by red pine root systems.

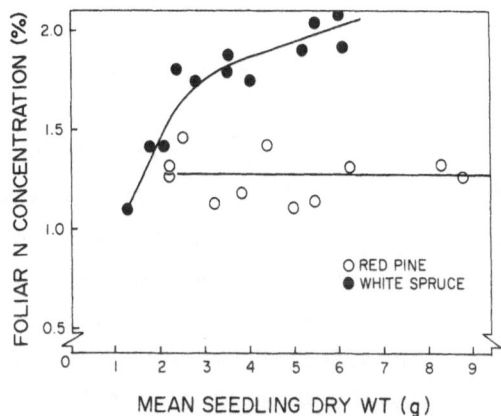

FIGURE 3. Foliage concentrations of nitrogen (N) in white spruce (<u>Picea glauca</u> [Moench] Voss) and red pine (<u>Pinus resinosa</u> Ait.) in relation to total seedling dry weight. (Adapted from [5])

Despite this example of red pine, fertilization can alter nutrient concentrations in seedling foliage. The extent of the alteration may depend on the species involved and on the influence of the other cultural practices yet to be discussed.

5.3.1.2 <u>Seedbed density</u>. Like fertilization, seedbed density is an easily regulated, efficient method for changing seedling nutrient concentration and morphology. Numerous studies have shown that as the seedling spacing increases, seedling dimensions increase (8, 9, 10, 14, 15, 48, 54, 55, 56, 70, 73, 75, 79, 98). The most notable effects of increased spacing are the increase in seedling dry weight, altered root morphology, and larger stem diameter. Height, however, usually is least affected. Indeed, McClain (46) illustrated that a decrease in spacing of 2+0 white spruce from 430 to 215 seedlings/m^2 produced the following increases: seedling dry weight (84%), root area (61%), stem diameter (44%), and height (22%). Larger crowns and an increased amount of foliage were also observed among the more widely spaced white spruce.

Seedbed density has an indirect effect on nutrient concentrations: decreasing the number of trees per unit area increases the nutrients available per seedling. For example, Bell (9) clearly showed that N concentration of the foliage increased as seedbed density decreased; his results included findings that seedbed densities of 788, 394, 197, 98 seedlings/m^2 resulted in foliar N concentrations of 1.46%, 1.66%, 1.91%, and 1.99%, respectively. A similar response trend was also reported by McClain and Armson (49), who also showed that nutritional differences due to seedbed density can be further affected by soil water supply.

Clearly, therefore, the control of seedbed density results in multiple changes in the seedlings. Most often, low seedbed densities alter seedling morphology, produce larger seedlings, and improve seedlings' nutritional status (particularly N concentration).

5.3.1.3 Irrigation. Regulation of irrigation is important in affecting the growth of nursery stock. Few studies, however, have addressed the influence of irrigation on plant nutrition. Nutrient movement in soil (particularly of P) is dependent on water movement (51); therefore, it is reasonable to assume that absorption of nutrients by plant root systems can be facilitated by using irrigation to manage the water content of nursery soils. Insufficient irrigation can retard growth, in part because the rate of nutrient absorption is reduced. Water deficiencies can restrict the elongation and development of new roots (20), thereby reducing the root surface area available for nutrient absorption.

The evaluation of soil water availability is difficult under field conditions, and has in many cases led to the use of greenhouse studies to develop relationships between seedling response and irrigation regimes. Schomaker (71) applied irrigation at varying frequencies and nutrient solutions of varying concentrations to white pine seedlings.

Seedling dry weights increased as the concentration of
nutrient solutions and frequency of irrigation increased.
However, seedlings given the highest strength nutrient solution showed increased foliar concentrations of N, P, K,
calcium (Ca), and magnesium (Mg) as the frequency of irrigation decreased (from every 7 to every 28 days).
Consideration of seedling dry weights revealed that, along
with increased foliar concentrations, seedling dry weight
decreased from 7.89 g to 3.51 g. Obviously, rapid growth of
frequently irrigated seedlings resulted in the dilution of
foliar nutrient concentrations and the slower growth of less
frequently irrigated seedlings resulted in the high nutrient
concentrations. In contrast, nutrient contents per seedling
decreased from a high of 34.8 mg with frequent irrigation to
22.9 mg with the least frequent irrigation regime. The
implication is that the influence of irrigation on foliar
nutrient content is mediated by seedling dry weight.

Similar findings were produced by another nursery study
(49) which examined the effects of irrigation on the growth
and nutrition of white spruce seedlings. The results indicated that with frequent irrigation N concentrations in the
foliage of seedlings were lower but N contents were higher
than for seedlings grown under conditions of natural precipitation only (no irrigation).

The effect of irrigation on the nutrition of plants is
most marked when nutrient levels are sufficiently high to
allow for adequate growth. At low levels of nutrition, the
benefits of irrigation are not realized.

5.3.1.4 Wrenching. Wrenching is a practice that is commonly employed to reduce the growth rate of rapidly growing
seedlings. Wrenching involves severing the lower part of
the root system and lifting the seedling up, but not out of,
the soil. This practice temporarily impedes the seedling's
absorption of water and nutrients, effectively reduces shoot
growth, and may lead later to the development of a fibrous
root system (65).

The effects of wrenching on seedling nutrition can be

described as transitory and related to induced morphological changes. As expected, nutrient absorption after wrenching decreases, as does absorption of water (23). van den Driessche (90) reported that, in most cases, wrenching of Douglas-fir seedlings had a greater effect on P and K uptake than on N uptake. The exceptions were wrenching treatments applied throughout the summer and late summer, which had equivalent effects on N, P, and K uptake.

Fertilization can lessen the effect of wrenching on nutrient uptake; in fact, a combined fertilization-wrenching treatment may actually increase nutrient concentrations in seedlings (90).

5.3.1.5 Interactive effects of nursery treatment on nutrient supply. Thus far we have considered only individual cultural practices and their effects on seedling nutrient concentrations. We have noted that nutrient concentrations generally rise when irrigation or the rate of fertilization increase, or when seedbed density decreases. Rarely, however, do nutrient concentrations vary within the plant in response to a single factor. Typically, the nutrient status of a plant or seedling is a reflection of the combined effects of more than one cultural practice, such as seedbed density, fertilization, and irrigation.

For example, Bell (9) determined the nutrient contents of white spruce seedlings in relation to seedbed density and fertility and demonstrated that N contents of the seedlings increased as fertilization increased for any of the four seedbed densities involved. The increase in foliar N, however, was greatest (64%) at the highest density level (860 seedlings/m^2). Similar results have been reported by Armson (5) for white spruce. Switzer and Nelson (79) evaluated the change in N concentration in relation to the level of fertilization at four seedbed densities, and demonstrated that the greatest increase in N concentration occurred between the medium and highest level of fertilization in the seedbed of the greatest density.

It should be recognized, however, that despite marked

increases in N concentrations at higher seedbed densities, the size of seedlings typically is depressed. Maximum seedling dry weight usually is realized at low seedbed densities, where seedlings also possess the highest nutrient contents. Higher fertilization rates at low seedbed density apparently provide abundant mineral nutrients; in these cases, growth may be restricted only by other limiting factors, such as water.

5.3.2 Nutrition and field performance

We have drawn attention to a number of studies that clearly demonstrate that seedling nutrition, as well as morphology, can be significantly altered by changing the intensity of a single cultural practice or combination of practices. Three of the most intensively studied cultural practices are discussed below.

5.3.2.1 Fertilization. Increasing the rate of fertilization in the nursery, and hence the nutrient content of the seedling, can result in both positive and negative effects after planting.

Positive effects of fertilization. Seedlings grown at moderately high fertilization levels in the nursery exhibit higher survival and rates of growth than seedlings grown at lower fertilization levels. Various authors have attributed this enhanced survival and growth to morphological attributes such as shoot:root ratio, stem diameter, seedling dry weight, root area, and leaf surface area, and to physiological attributes such as nutritional status, carbohydrate reserves, dormancy and drought resistance. A brief listing of results of various investigations is presented below.

- White spruce survival and height growth were increased by fertilization, and N concentration in 2+0 foliage was positively correlated to height growth (9).
- White spruce survival and growth after 5 years were increased by fertilization, although there was some variability, depending on planting site. Aggregate

height was increased 15% at the highest fertilization rate (55).

- Fall fertilization of Douglas-fir significantly increased fifth-year survival over control (88% vs. 81%); fertilized trees were also taller than control trees (4).

- Loblolly pine (<u>Pinus taeda</u> L.) survival, regardless of fertility level, varied between 92% and 96% after 3 years. Height of seedlings at 3 years, however, was positively related to fertilizer level and plant nutrition (Figure 4) (79).

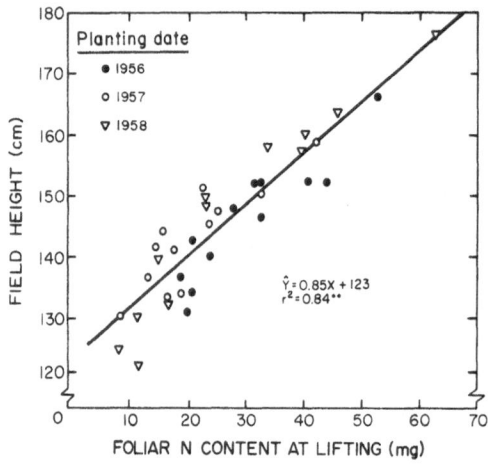

FIGURE 4. The relationship between seedling height 3 years after planting to foliar nitrogen (N) concentrations at time of lifting. (Adapted from [79])

- Douglas-fir survival (Figure 5) and height 2 years after outplanting were significantly affected by N application in the nursery (87).

- Highest N supply level in the nursery increased survival at 3 years of coastal Douglas-fir (<u>Pseudotsuga menziesii</u> [Mirb.] Franco var. menziesii) and Sitka spruce, but did not affect survival of lodgepole pine (at 2 years) or of interior Douglas-fir (<u>Pseudotsuga menziesii</u> var. glauca [Beissh.] Franco). Three years after planting, new shoot growth at the highest N level ranged from no change to 42% greater than growth at the lowest N level (89).

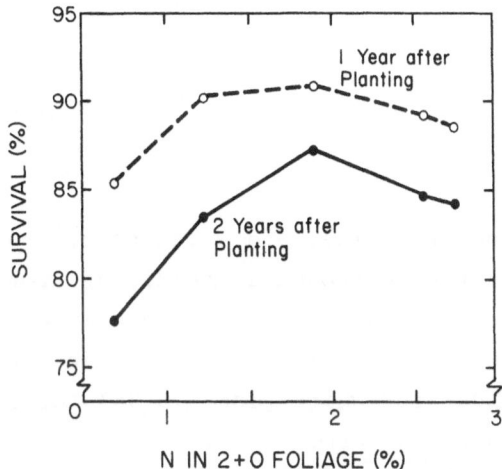

FIGURE 5. Average survival of 2+0 Douglas-fir (<u>Pseudotsuga menziesii</u> [Mirb.] Franco) seedlings 1 and 2 years after planting in relation to foliar nitrogen (N) concentration at lifting. Points represent five levels of N applied in nursery. (Adapted from [87])

<u>Negative effects of fertilization</u>. Negative effects of fertilization have been associated with reduced survival and growth, frost damage, and reduced drought resistance.

- Fall application of N to Sitka spruce advanced bud break the following spring and resulted in frost damage (11).
- Lodgepole pine seedlings grown at high nitrate levels exhibited low survival when subjected to drought (25).
- White spruce seedlings grown at the highest N supply level in the nursery exhibited the lowest survival 3 years after planting. Growth of seedlings was least for seedlings grown at the highest N level (47).
- Loblolly pine seedlings grown at above excessive N levels exhibited reduced drought resistance (59).
- Loblolly, longleaf (<u>Pinus palustris</u> Mill.) and slash pines (<u>Pinus elliotti</u> Engelm.) receiving high N applica-

tions in the fall experienced reduced first-year field survival (71).

5.3.2.2 Seedbed density. The effects of seedbed density have been noted to affect seedling morphology as well as nutritional status. Almost without exception, seedlings grown at low seedbed densities exhibited higher survival and growth rates than did seedlings grown at high seedbed densities.

For example:

- White pine, red pine, and white spruce survival was significantly increased in seedlings grown at low seedbed densities. As seedbed densities decreased from 431 to 108 seedlings/m^2, increases in survival ranged from 65-100% for white pine, 75-95% for red pine and 70-95% in white spruce (15).

- Radiata pine was grown at varying spacings between and within drill rows, creating different densities. Two years after outplanting, tree size and survival were significantly reduced for plants raised at the high seedbed densities (10).

- Douglas-fir was grown at five seedbed densities; after four growing seasons, survival ranged from 70% to 80%. Survival was not affected, but seedling height was significantly affected by density (36).

5.3.2.3 Irrigation. No report on the influence of irrigation on seedling nutrition and subsequent growth and survival could be located.

5.3.2.4 Wrenching. The impact of wrenching on seedling field performance presently is not well defined. Results from wrenching trials in New Zealand (65) demonstrate the success of this technique in preparing radiata pine for outplanting; however, in the Pacific Northwest, success of wrenched Douglas-fir is variable (23, 90).

van den Driessche (90) examined survival and growth in relation to nutrition and presented data indicating that differences in the timing and severity of wrenching produced little variation in survival (41% to 49%) two seasons after

planting. The wrenching treatments also did not affect N
concentrations and had little effect on second-year new
shoot growth.

5.4 CARBOHYDRATE RESERVES

Carbohydrates are an extremely important component of
trees, making up 75% of their total dry weight (37).
Carbohydrates are classified into three major groups: mono-
saccharides (for example, glucose and fructose), disacchari-
des (for example, sucrose and maltose), and polysaccharides
(for example, cellulose and starch).

Carbohydrates produced by photosynthesis are for the most
part either oxidized in respiration or used to synthesize
other important organic compounds and structural parts of
trees. Trees store carbohydrates when the rate of photosynthetic production of carbohydrates exceeds the rate of
use (37). Tree seedlings primarily store food reserves in
the form of starch and sucrose, but also in the form of
hemicelluloses, fats, proteins, and other compounds (37,
62).

The primary use of carbohydrate reserves is to maintain
respiration and growth during times when photosynthates are
not produced by photosynthesis (37). Most reserves are
accumulated during the latter part of the growing season
and, to varying degrees (depending on the environment and
species) during the fall and winter. These reserves are
then used for plant maintenance in the winter, and for initial growth the following growing season (2, 30, 39, 40).
Both spring root growth and shoot elongation have been shown
to depend on carbohydrate reserves (e.g., 2, 30, 37, 38).

This section reviews some of the nursery practices that
affect carbohydrate reserve levels, and the effects that
these altered levels may have on field performance.

5.4.1 Nursery practices and carbohydrate reserves

5.4.1.1 Stressful growing conditions. Little information
is available on the effects of different nursery cultural

practices on carbohydrate reserves. It is believed, however, that under stressful, versus optimal, growing conditions, the storage of these foods is not promoted (21, 23, 27, 32). Cultural practices or growing conditions that can cause stress, and therefore could result in a reduction in photosynthetic capacity and the accumulation of carbohydrates, include:

- An irrigation regime that imposes severe plant moisture stress (27, 35, 44, 58).
- Undercutting and root wrenching that result in high plant moisture stress (23, 45).
- Transplanting in the nursery or into holding beds, which stresses seedlings (64).
- High seedbed density that reduces resources (for example, light, water, and nutrients) available for optimum photosynthesis (9, 49).
- Nutrient deficiencies that result in a decrease in growth or in a disruption of physiological processes (49, 79).
- Seedling insect and disease outbreaks that cause defoliation, reduction in photosynthetic capacity and disruption of other physiological processes (58, 94).
- Lifting seedlings too early in the fall, when they are not as resistant to handling and have not had a chance to maximize their carbohydrate reserve levels (40, 41, 45).

5.4.1.2 Cold storage. The depletion of carbohydrate reserves in dark, cold storage has been investigated in Douglas-fir (61, 99), ponderosa pine (Pinus ponderosa Dougl. ex Laws) (32), mugo pine (Pinus mugo Terra) and radiata pine (50), Englemann spruce (Picea engelmannii) (64), and Scots pine (Pinus sylvestris L.) (60). Most of these studies have shown reserve levels to decrease with cold storage, but some have not. Reasons for these differences in results could include the effect of lifting date, storage temperature, length of storage, and other cultural practices that result in varied carbohydrate levels in seedlings.

Results of some of these cold storage studies are summarized below:

- An early study reported that storage of Douglas-fir and
  noble fir for 1 month did not affect carbohydrate levels
  in the foliage (99); however, a more recent study showed
  that carbohydrates decreased rapidly in the foliage,
  stem, and roots during storage at 2°C and -1°C (Figure
  6) (61). This study showed that carbohydrate levels
  declined more rapidly in cold-stored (at 2°C) seedlings
  than in freezer-stored (-1°C) ones; freezer-stored
  seedlings took 6 months to drop to the reserve levels
  reached by cold-stored seedlings in only 2 months.

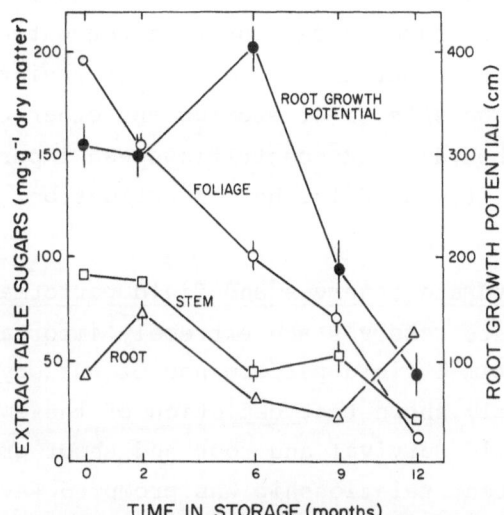

FIGURE 6. Carbohydrate (extractable sugar) levels in
foliage, stems, and roots and root growth potential (RGP) of
2+0 Douglas-fir (Pseudotsuga menziesii [Mirb.] Franco)
seedlings stored for up to 12 months at -1°C. (Vertical
lines indicate standard error; absence of vertical lines
indicates standard error is near 0.) (Adapted from [61])

- Carbohydrate levels of Engelmann spruce seedlings
  decreased when the seedlings were stored at 0°C for 4
  months; the most rapid decline occurred during the last
  2 months of storage (64).
- Carbohydrate concentration of both mugo and radiata

pines declined steadily during storage that lasted up to 18 weeks (50). McCracken (50) noted that radiata pine had lower concentrations in its stems and roots than did mugo pine and that radiata pine, in turn, does not store as well as mugo pine.

Cold storage is an accepted and necessary practice for holding forest tree seedlings between the time they are lifted, and planted. The storage period can vary, from as short as 1 week, to as long as 6 months for seedlings that will be planted at higher elevations. There are marked differences in field performance following storage, and these differences could be due to the effect of storage on carbohydrate reserves. Ritchie (61) suggests that the influence of storage on carbohydrate concentrations may affect both frost hardiness and drought resistance of seedlings. The effects of storage and other cultural practices on carbohydrate concentrations--an important aspect of seedling quality--need further investigation.

5.4.2 Carbohydrate reserves and field performance

Carbohydrate reserves are extremely important in achieving optimal field performance of nursery stock; it has been repeatedly shown that depletion of these reserves decreases field survival and root and shoot growth.

This apparent relationship has prompted several authors to suggest that carbohydrate reserve levels be used as indicators of seedling quality or physiological condition (32, 60, 62, 94); however, no single investigation has soundly established this relationship.

5.4.2.1 Survival. Poor seedling survival has been attributed to depleted carbohydrate reserve levels during cold storage in Jeffrey pine (Pinus jeffrey Grev. & Balf.) and ponderosa pine (32), Scots pine (60), several Appalachian hardwoods (27), white spruce (34), and radiata and mugo pines (50). Winjum (99) found no change in carbohydrates in Douglas-fir seedlings stored for 4 weeks; however, the highest field survival was found for seedlings lifted in mid-January, which had the highest levels of non-reducing sugars.

Two authors (60, 64) have suggested that there is a critical threshold for carbohydrate reserve levels, and that a seedling is adversely affected when its level falls below a particular concentration. Puttonen (60) noted that mortality of Scots pine seedlings increased when total glucose content of the needles fell below 2% of needle dry weight. Ronco (64) reported that first-year survival of Engelmann spruce planted in the Rocky Mountains seemed especially dependent on carbohydrate reserves because stressful site conditions immediately after planting tended to favor depletion, rather than accumulation, of reserves.

It has also been suggested that adequate carbohydrate reserves increase stress resistance of seedlings (60, 61). Ritchie (61) noted that foliar sugar levels are highly correlated with frost hardiness of Douglas-fir seedlings, and that the depletion of sugars during cold storage may reduce drought resistance in seedlings.

A study of large Douglas-fir and noble fir trees that had been defoliated by insects reported that trees with a detectable level of carbohydrate reserves (in this case, starch) recovered, whereas trees lacking detectable carbohydrate reserves died (94). Webb's (94) main point was that defoliation stressed trees, causing depletion of carbohydrate reserve levels; these trees, when later exposed to stressful environmental conditions, did not survive. This depletion may resemble the depletion that occurs in seedlings that undergo prolonged cold storage, which subsequently reduces their survival potential.

5.4.2.2 Shoot growth. Seedling shoot growth in the spring depends largely on carbohydrate reserves (30, 38, 39). These reserves may be translocated primarily from old needles, whose dry-weight decrease sometimes can roughly equal the new shoots' dry-weight increase (38). Large amounts of these reserves are also used by the high rate of respiration of the expanding shoot (38).

The importance of carbohydrate reserves to spring shoot
growth varies and probably is most affected by pre-lifting,
post-lifting, and planting environments. One study (30)
reported that 1-year-old needles of red pine seedlings
supplied reserves for shoot elongation in three ways: 1) by
translocating photosynthates to the bud during the previous
year's growing season, 2) by meeting respiration demands of
the shoot, and 3) by providing current photosynthates during
spring shoot growth. Webb (94) demonstrated that Douglas-
fir saplings defoliated by insects had greater needle length
and shoot regrowth in the spring, when starch content was
greater in the tree stems; in this study, starch content
apparently was a good indicator of the trees' ability to
grow after being subjected to stressful conditions. Scots
pine seedlings with low glucose content of needles at the
time of field planting had shorter needles and greater mortality of the terminal shoots at the end of the growing
season than did seedlings whose needles had a higher glucose
content (60).

Multiple tops are also a concern in reforestation.
Kozlowski and Winget (38) suggested that a supply of carbohydrate reserves from the leaves appears necessary in
maintaining dominance of the terminal leader in red pine
seedlings. They note that the degree of apical dominance
(meaning fewer seedlings with multiple tops) depends on the
availability of N and carbohydrates to the growing shoot.

5.4.2.3 Root growth. After being planted, seedlings must
rapidly regenerate new roots to establish close contact with
the soil and satisfy the water demands of the shoots (63,
77). This ability to regenerate new roots, often called
root growth capacity or root growth potential, is closely
linked to the seedling's ability to avoid water stress after
planting (57, 77). Root growth potential is therefore a
critical factor in reforestation success.

The few studies that have addressed the question of the
importance of carbohydrate reserves to root growth potential
have presented conflicting results. In one study (74),

3-year-old white pine seedlings exposed to $^{14}CO_2$ were found
to have high levels of ^{14}C translocated to the roots in the
spring when root activity was the greatest; the authors
suggested that the increased activity of the roots caused
carbohydrates to move to the roots. Most studies that
emphasize the importance of carbohydrates to root growth
have shown that carbohydrate levels in the roots are high
during periods of increased root growth (40, 74). These
studies, however, have not shown that there is a direct
cause-and-effect relationship between carbohydrate reserves
and root growth.

Root growth is an energy-consuming process that requires
metabolic substrates, which are mainly carbohydrates.
However, an important question about root growth remains
unanswered: Does new root growth that occurs after planting
use carbohydrate reserves, or carbohydrates currently produced by photosynthesis, or both? Ritchie and Dunlap's
excellent review (63) of this question notes that, although
carbohydrates are required for root growth, the level of
carbohydrate reserves is not the sole influence on root
growth. Furthermore, in a more recent study Ritchie (61)
reported that carbohydrates generally declined steadily
during cold storage but that root growth potential increased
during the first 6 months of storage (Figure 6). Other
investigators have also been unable to find correlations
between root growth potential and carbohydrate reserves (26,
86). Hence, Ritchie (61) has proposed some alternative physiological factors that might control root growth potential;
those factors are the supply of current photosynthate from
the shoot and/or the status of bud dormancy.

5.5 CONCLUSIONS AND RECOMMENDATIONS

5.5.1 Frost hardiness

Frost hardiness can be defined as the minimum temperature
to which a tree seedling can be exposed without being
damaged. Seedlings may be frost damaged in the nursery or

after field planting. Damage can range from slight needle damage to top dieback to seedling mortality.

Timing of the irrigation and fertilizer regimes in a nursery can have a tremendous impact on frost hardiness. A moderate moisture stress regime can induce dormancy and frost hardiness in Douglas-fir seedlings, whereas unrestricted watering can extend the growing season and result in frost damage to seedlings. Fertilization (especially with N) can prolong growth, delay the onset of dormancy and result in frost damage to succulent shoots. However, fertilization that does not extend the growing season has been shown to increase frost hardiness. Species, seed source and nursery location can also influence the development of frost hardiness.

Frost hardiness can affect field performance in three ways: 1) seedlings that have sustained either detectable or undetectable damage in the nursery may be planted in the field, 2) seedlings may be lifted, processed, and stored or field planted when they are not frost hardy, and 3) field-planted seedlings that burst bud early in the spring may be subjected to subfreezing temperatures. We stress the importance of preventing frost damage in the nursery, and recommend that practices that promote induction of frost hardiness and enable seedlings to tolerate lifting, handling, storage and planting should be implemented at nurseries. Nursery managers should also assess the frost hardiness of seedlings that will be planted on sites where cold temperatures are expected to be a problem.

5.5.2 Mineral nutrition

The importance of mineral nutrition to outplanting performance has been exemplified by the cited literature. Deficiencies of one or more micronutrients or macronutrients can impair nursery seedling growth, retard establishment and early rapid growth, or lessen survival in the field. Achievement of the appropriate nutrient status, which may be species specific, can be attained by one, but more commonly

by a combination, of cultural practices. Increased nutrient
concentrations typically are associated with higher survival
and greater shoot growth; variations in this general relationship probably are due to the interaction of the site
with seedling physiology. Production of planting stock with
nutritional levels characteristic of rapidly growing
seedlings requires periodic assessment of nutrient status
with the aim to correct deficiencies. Nursery managers,
however, must not undertake corrective activities that will
alter seedling proportions and/or other physiological
attributes, such as frost hardiness and carbohydrate levels.

It probably is impossible to define a specific nutrient
concentration for all species, on all sites. Present indi-
cations are that survival and growth is enhanced in
seedlings whose foliar N concentrations range from 1.7% to
2.3%. We urge nursery managers, however, to be cautious in
their efforts to achieve excessively high nutrient concentrations in the hope of attaining exceptional
performance; such concentrations rarely result in that performance.

5.5.3 Carbohydrate reserves

The primary use of carbohydrate reserves is to maintain
respiration and growth when photosynthates are not produced
by photosynthesis. Any nursery cultural practice (such as
irrigation regime, root wrenching, and high seedbed density)
that causes stress may result in a reduction of seedling
photosynthetic capacity and accumulation of carbohydrate
reserves. Cold storage, an accepted and necessary practice
for holding seedlings between lifting and planting, causes
carbohydrate reserves to decrease; poor seedling survival in
a number of species has been attributed to carbohydrate
reserve depletion during cold storage. Depletion of carbohydrate reserves may markedly affect first-year growth as
well, because a seedling's initial root and shoot growth
after field planting may also depend largely on carbohydrate
reserves. There are no definitive studies of the cause-and-

effect relationships of nursery practices, carbohydrate
levels, and field performance; however, careful attention
should be paid to those practices that may deplete carbohydrate reserves and lessen chances for optimal field performance. Cold and freezer storage should be further
investigated to better define their effects.

REFERENCES

1. Alden, J., and R. K. Hermann. 1971. Aspects of cold
    hardiness mechanisms in plants. Bot Rev 37:37-142.
2. Allen, R. M. 1964. Contribution of roots, steams,
    and leaves to height growth of longleaf pine. Forest
    Sci 10:14-16.
3. Allison, C. J. 1972. Freeze-damage control in lowest
    nurseries. Proc Int Plant Propagations Soc 22:77-82.
4. Anderson, H. W., and S. P. Gessel. 1966. Effects of
    nursery fertilization on outplanted Douglas-fir. J
    For 64:109-112.
5. Armson, K. A. 1968. The effects of fertilization and
    seedbed density on the growth and nutrient content of
    white spruce and red pine seedlings. Univ Toronto.
    Tech Rep 10. 16 p.
6. Armson, K. A., and V. Sadreika. 1979. Forest tree
    nursery soil management and related practices.
    Ontario Minist Nat Resour. 179 p.
7. Arnold, C. I. 1958. Frost damage. USDA Forest Serv
    Tree Planters' Notes 31:8-9.
8. Baron, F. J., and G. H. Schubert. 1963. Seedbed den-
    sity and pine seedling graders in California nur-
    series. USDA Forest Serv Res Note PSW-31, Pac
    Northwest Forest and Range Exp Stn, Portland, Oreg.
9. Bell, T. I. W. 1968. Effect of fertilizer and den-
    sity pretreatment on spruce seedling survival and
    growth. Forest Record 67:1-67. For Comm, Grt Britain.
10. Benson, A. D., and K. R. Shepherd. 1976. The effect
    of nursery practice on Pinus radiata seedling charac-
    teristics and field performance. N Z J For Sci
    6:19-26.
11. Benzian, B., R. M. Brown, and S. C. R. Freeman. 1974.
    Effect of late season top dressing of N (and K)
    applied to conifer transplants in the nursery on their
    survival and growth on British forest sites. Forestry
    47:153-184.
12. Blake, J., J. B. Zaerr, and S. Hee. 1979. Controlled
    moisture stress to improve cold hardiness and morpho-
    logy of Douglas-fir seedlings. Forest Sci 25:576-582.
13. Brown, G. N. 1976. Cold hardiness in tree species.
    In: Proc 11th IUFRO World Cong, Div II, Norway, p.
    46-49.

14. Bunting, W. R. 1973. Seedbed density trials, white pine, red pine and white spruce, Orono Nursery. Ontario Minist Nat Resour Nursery Notes No 34.

15. Bunting, W. R. 1980. Seedling quality: Growth and development - soil relationships, seedling growth and development, density control relationships. In: Proc North Am Forest Tree Nursery Soils Workshop (L. P. Abrahamson and D. H. Bichelhaupt, eds), p. 21-42. SUNY College, New York.

16. Burdett, A. N. 1983. Quality control in the production of forest planting stock. For Chron 59(3):132-138.

17. Burke, M. J., L. V. Gusta, H. A. Quamme, C. J. Weiser, and P. H. Li. 1976. Freezing and injury in plants. Annu Rev Plant Physiol 27:507-528.

18. Campbell, R. K., and F. C. Sorensen. 1973. Cold-acclimation in seedling Douglas-fir related to phenology and provenance. Ecology 54:1148-1151.

19. Clarkson, D. T., and J. B. Hanson. 1980. The mineral nutrition of higher plants. Annu Rev Plant Physiol 31:239-298.

20. Day, R. J., and G. R. MacGillivrary. 1975. Root regeneration of fall-lifted white spruce nursery stock in relation to soil moisture. For Chron 51:196-199.

21. Duryea, M. L. 1984. Nursery cultural practices: impacts on seedling quality. In: Forest nursery manual: Production of bareroot seedlings (M. L. Duryea and T. D. Landis, eds), p. 143-164. Martinus Nijhoff/Dr W. Junk Publishers, The Hague/Boston/Lancaster, for Forest Res Lab, Oreg State Univ.

22. Duryea, M. L., and T. D. Landis. 1984. Development of the forest nursery manual: A synthesis of current practices and research. In: Forest nursery manual: Production of bareroot seedlings (M. L. Duryea and T. D. Landis, eds), p. 3-8. Martinus Nijhoff/Dr W. Junk Publishers, The Hague/Boston/Lancaster, for Forest Res Lab Oreg State Univ.

23. Duryea, M. L., and D. P. Lavender. 1982. Water relations, growth, and survival of root wrenched Douglas-fir seedlings. Can J Forest Res 12:545-555.

24. Edgren, J. W. 1970. Growth of frost-damaged Douglas-fir seedlings. USDA Forest Serv Res Note PNW-121, 8 p. Pac Northwest Forest and Range Exp Stn, Portland, Oreg.

25. Etter, H. M. 1969. Growth, metabolic components and drought survival of lodgepole pine seedlings at three nitrate levels. Can J Plant Sci 49:393-402.

26. Etter, H. M., and W. L. Carlson. 1973. Sugars, relative water content, and growth after planting of dormant lodgepole pine seedlings. Can J Plant Sci 53:395-399.

27. Farmer, R. E., Jr. 1975. Seasonal carbohydrate levels in roots of Appalachian hardwood planting stock. USDA Forest Serv Tree Planters' Notesing 29:22-24.

28. Giertych, M., and H. Faber. 1967. Variation among Norway spruce of Polish provenance in seedling growth and nitrogen uptake. In: Proc 14th IUFRO World Cong, Div III, p. 536-550.

29. Glerum, C. 1976. Frost hardiness of forest trees. In: Tree physiology and yield improvement (M. G. R. Cannell and F. T. Last, eds), p. 403-420. Academic Press, New York.

30. Gordon, J. C., and P. R. Larson. 1970. Redistribution of ^{14}C-labeled reserve food in young red pines during shoot elongation. Forest Sci 16:14-20.

31. Griffin, A. R. 1974. Geographic variation in juvenile growth characteristics of Douglas-fir [Pseudotsuga menziesii (Mirb.) Franco.] from the coastal ranges of California. Ph.D. thesis, College of For, Oregon State Univ, Corvallis. 153 p.

32. Hellmers, H. 1962. Physiological changes in stored pine seedlings. USDA Forest Serv Tree Planters' Notes 53:9-10.

33. Hocking, D., and R. D. Nyland. 1971. Cold storage of coniferous seedlings: A review. Appl For Res Inst Rep No 6, 70 p. State Univ Coll For, Syracuse, New York.

34. Hocking, D., and B. Ward. 1972. Late lifting and freezing in plastic bags improve white spruce survival after storage. USDA Forest Serv Tree Planters' Notes 23:24-26.

35. Hsiao, T. C., E. Acevedo, E. Fereres, and D. H. Henderson. 1976. Stress metabolism: Water stress, growth, and osmotic adjustment. Philos Trans R Soc London Ser B 273, 479-500.

36. Iverson, R. D. 1981. Low seedbed densities enhance Douglas-fir seedling size and performance. Tech Note No 60. Forest Prod Res, Int Paper Co, Oreg.

37. Kramer, P. J., and T. T. Kozlowski. 1979. Physiology of woody plants. Academic Press, New York. 811 p.

38. Kozlowski, T. T., and C. H. Winget. 1964. The role of reserves in leaves, branches, stems, and roots on shoot growth of red pine. Amer J Bot 51:522-529.

39. Krueger, K. W. 1967. Nitrogen, phosphorus, and carbohydrates in expanding and year-old Douglas-fir shoots. Forest Sci 13:352-356.

40. Krueger, K. W., and J. W. Trappe. 1967. Food reserves and seasonal growth of Douglas-fir seedlings. Forest Sci 13:192-202.

41. Lavender, D. P. 1964. Date of lifting for survival of Douglas-fir seedlings. Oreg State Univ Forest Res Lab Res Note 49.

42. Lavender, D. P., K. K. Ching, and R. K. Hermann. 1968. Effect of environment on development of dormancy and growth of Douglas-fir seedlings. Bot Gaz 12:70-83.

43. Levitt, J. 1956. The hardiness of plants. Academic Press, New York. 278 p.

44. Levitt, J. 1972. Responses of plants to environmental stresses. Academic Press, New York. 697 p.

45. Little, C. H. A. 1970. Derivation of springtime starch increase in balsam fir (<u>Abies balsamea</u>). Can J Bot 48:1995-1999.

46. McClain, K. M. 1977. Differential effects of seedbed density on seedling growth. Ontario Minist Nat Resour Forest Res Note No 7.

47. McClain, K. M. 1983. Seedbed spacing and nitrogen effects on nursery growth and post planting performance of white and black spruce. <u>In</u>: Proc Nurserymen's Meet, June 7-11, 1982. Thunder Bay, Ontario. Ontario Minist Nat Resour.

48. McClain, K. M., and K. A. Armson. 1975. Growth responses of pine and spruce seedlings to regimes of soil moisture and fertility. Soil Sci Soc Am Proc 39:140-144.

49. McClain, K. M., and K. A. Armson. 1976. Effect of water supply, nitrogen and seedbed density on white spruce seedling growth. Soil Sci Soc Am J 40:443-446.

50. McCracken, I. J. 1979. Changes in the carbohydrate concentration of pine seedlings after cool storage. N Z J Forest Sci 9:34-43.

51. Mengel, K., and E. A. Kirkby. 1982. Principles of plant nutrition. 3rd ed. Int Potash Inst, Bern, Switzerland. 655 p.

52. Menzies, M. I., D. G. Holden, L. M. Green, and D. A. Rook. 1981. Seasonal changes in frost tolerance of <u>Pinus radiata</u> seedlings raised in different nurseries. N Z J Forest Sci 11:100-111.

53. Morrison, I. K. 1974. Mineral nutrition of conifers with special reference to nutrient status interpretation: A review of literature. Dep Environ, Can Forest Serv Publ No 1343. 74 p.

54. Mullin, R. E., and L. Bowdery. 1977. Effects of seedbed density and nursery fertilization on survival and growth of 3+0 white pine. USDA Forest Serv Tree Planters' Notes 28:11-13.

55. Mullin, R. E., and L. Bowdery. 1977. Effects of seedbed density and nursery fertilization on survival and growth of white spruce. For Chron 53:83-86.

56. Mullin, R. E., and L. Bowdery. 1978. Effects of seedbed density and top dressing fertilization on survival and growth of 3+0 red pine. Can J Forest Res 8:30-35.

57. Nambiar, E. K. S., G. D. Bowen, and R. Sands. 1979. Root regeneration and plant water status of <u>Pinus radiata</u> D. Don seedlings transplanted to different soil temperatures. J Exp Bot 30:1119-1131.

58. Parker, J., and R. L. Patton. 1975. Effects of drought and defoliation on some metabolites in roots of black oak seedlings. Can J Forest Res 5:457-463.

59. Pharis, R. P., and P. J. Kramer. 1964. The effect of nitrogen and drought on loblolly pine seedlings. Forest Sci 10:143-150.

60. Puttonen, P. 1980. Effect of temporary storage tem-
     perature on carbohydrate levels in Scots pine
     seedlings and planting success. In: Characterization
     of plant material. Int Meet IUFRO, Group S1.05 04.
     June 23-29, 1980 Frieburg.

61. Ritchie, G. A. 1982. Carbohydrate reserves and root
     growth potential in Douglas-fir seedlings before and
     after cold storage. Can J Forest Res 12:905-912.

62. Ritchie, G. A. 1984. Assessing seedling quality.
     In: Forest nursery manual: Production of bareroot
     seedlings. (M. L. Duryea and T. D. Landis, eds), p.
     243-259. Martinus Nijhoff/Dr W. Junk Publishers, for
     Forest Res Lab, Oreg State Univ.

63. Ritchie, G. A., and J. R. Dunlap. 1980. Root growth
     potential: Its development and expression in forest
     tree seedlings. N Z J Forest Sci 10:218-248.

64. Ronco, F. 1973. Food reserves of Engelmann spruce
     planting stock. Forest Sci 19:213-219.

65. Rook, D. A. 1971. Effect of undercutting and
     wrenching on growth of Pinus radiata D. Don seedlings.
     J Appl Ecol 8:477-490.

66. Rudolf, P. O. 1939. Why forest plantations fail. J
     For 37:377-383.

67. Sakai, A., and C. J. Weiser. 1973. Freezing
     resistance of trees in North America with reference to
     tree regions. Ecology 54:118-126.

68. Salisbury, F. B., and C. W. Ross. 1978. Plant
     physiology. Wadsworth Publishing Co, Inc. 422 p.

69. Sandvik, M. 1980. Environmental control of winter
     stress tolerance and growth potential in seedlings of
     Picea abies (L.) Karst. N Z J Forest Sci 10:97-104.

70. Scarborough, N. M., and R. M. Allen. 1954. Better
     longleaf seedlings from low density nursery beds.
     USDA Forest Serv Tree Planters' Notes 18:29-32.

71. Schomaker, C. E. 1969. Growth and foliar nutrition
     of white pine seedlings as influenced by simultaneous
     changes in moisture and nutrient supply. Soil Sci Soc
     Am Proc 33:614-618.

72. Schopmeyer, C. S. 1940. Survival in forest plan-
     tations in the northern Rocky Mountain region. J For
     38:16-24.

73. Shipman, R. D. 1964. Low seedbed densities can
     improve early height growth of planted slash and
     loblolly pine seedlings. J For 62:814-817.

74. Shiroya, T., G. R. Lister, V. Slankis, G. Krotkov, and
     C. D. Nelson. 1966. Seasonal changes in respiration,
     photosynthesis, and translocation of the ^{14}C labelled
     products of photosynthesis in young Pinus strobus L.
     plants. Ann Bot (London) 30:81-91.

75. Shoulders, E. 1961. Effect of nursery bed density on
     loblolly and slash pine seedlings. J For 59:576-579.

76. Sorensen, F. C. 1978. Date of sowing and nursery
     growth of provenance of Pseudotsuga menziesii given
     two fertilizer regimes. J Appl Ecol 15:273-279.

77. Stone, E. C. 1970. Variation in the root growth

capacity of ponderosa pine transplants. In: Regeneration of ponderosa pine (R. K. Hermann, ed), p. 40-46. Oreg State Univ, Corvallis.

78. Sutton, R. F. 1979. Planting stock quality and grading. Forest Ecol Manag 2:123-132.

79. Switzer, G. L., and L. E. Nelson. 1963. Effect of nursery fertility and density on seedling characteristics, yield, and field performance of loblolly pine, Pinus taeda. Soil Sci Soc Am Proc 27:461-464.

80. Tanaka, Y., and R. Timmis. 1974. Effects of container density on growth and cold hardines of Douglas-fir seedlings. In: Proc. North Am Containerized Forest Tree Seedling Symp. (R. W. Tinus, W. I. Stein, and W. E. Balmer, eds), p. 181-186. Great Plains Agric Counc Publ 68.

81. Thompson, B. 1983. Why fall fertilize? In: Proc West Nurserymen's Conf, West Forest Nursery Counc, Medford, Oreg, August 10-12, 1982, p. 85-91.

82. Timmis, R. 1974. Effect of nutrient stress on growth, bud set, and hardiness in Douglas-fir seedlings. In: Proc North Am Containerized Forest Tree Seedling Symp (R. W. Tinus, W. I. Stein, and W. E. Balmer, eds, p. 187-193. Great Plains Agric Counc Publ 68.

83. Timmis, R. 1976. Methods of screening tree seedlings for frost hardiness. In: Tree physiology and yield improvement (M. G. R. Cannell and F. T. Last, eds), p. 421-435. Academic Press, London.

84. van den Driessche, R. 1969. Influence of moisture supply, temperature, and light on frost hardiness changes in Douglas-fir seedlings. Can J Bot 47:1765-1772.

85. van den Driessche, R. 1970. Influence of light intensity and photoperiod on frost-hardiness development in Douglas-fir seedlings. Can J Bot 48:2129-2134.

86. van den Driessche, R. 1978. Seasonal changes in root growth capacity and carbohydrates in red pine and white spruce nursery seedlings. In: Proc IUFRO Symp on Root Physiology and Symbiosis, Sept 11-15, Nancy, France (A. Riedacker and J. Gagnaine-Michard, eds), p. 6-19.

87. van den Driessche, R. 1980. Effect of nitrogen and phosphorus fertilization on Douglas-fir nursery growth and survival after out-planting. Can J Forest Res 10:65-70.

88. van den Driessche, R. 1980. Health, vigour, and quality of conifer seedlings in relation to nursery soil fertility. In: Proc North Am forest tree nursery soils workshop (L. P. Abrahamson and H. Bickelhaupt, eds), p. 100-120. State Univ New York Coll Environ Sci For, Syracuse.

89. van den Driessche, R. 1982. Relationship between spacing and nitrogen fertilization of seedlings in the nursery, seedling size, and outplanting performance. Can J Forest Res 12:865-875.

114

90. van den Driessche, R. 1983. Growth, survival, and physiology of Douglas-fir seedlings following root wrenching and fertilization. Can J Forest Res 13:270-278.

91. Wakeley, P. C. 1948. Physiological grades of southern pine nursery stock. Proc Soc Am For 43:311-322.

92. Warrington, I. J., and D. A. Rook. 1980. Evaluation of techniques used in determining frost tolerance of forest planting stock: A review. N Z J Forest Sci 10:116-132.

93. Webb, W. L. 1975. Dynamics of photoassimilated carbon in Douglas-fir seedlings. Plant Physiol 56:455-459.

94. Webb, W. L. 1981. Relation of starch content to conifer mortality and growth loss after defoliation by the Douglas-fir tussock moth. Forest Sci 27:224-232.

95. Weiser, C. J. 1970. Cold resistance and injury in woody plants. Science 169:1269-1278.

96. Wilde, S. A. 1958. Forest Soils. Ronald Press, New York.

97. Williams, R. D., D. T. Funk, R. E. Phares, W. Lemmien, and T. E. Russell. 1974. Apparent freeze damage to black walnut seedlings related to seed source and fertilizer treatments. USDA Forest Serv Tree Planters' Notes 25:6-8.

98. Wilson, B. C. and R. K. Campbell. 1972. Seedbed density influences height, diameter, and dry weight of 3+0 Douglas-fir. USDA Forest Serv Tree Planters' Notes 23:1-4.

99. Winjum, J. K. 1963. Effects of lifting date and storage on 2+0 Douglas-fir and noble fir. J For 61:648-654.

100. Zaerr, J. B., B. D. Cleary, and J. Jenkinson. 1980. Scheduling irrigation to induce seedling dormancy. In: Proc Joint Meet Intermountain Nurserymen's Assoc and West Forest Nursery Counc, Boise, Idaho, p. 74-78.

6. SEED SOURCE LIFTING WINDOWS IMPROVE PLANTATION ESTABLISHMENT
OF PACIFIC SLOPE DOUGLAS-FIR

J. L. JENKINSON

Plant Physiologist, Pacific Southwest Forest and Range
Experiment Station, Forest Service, U.S. Department of
Agriculture, Berkeley, California 94701

ABSTRACT

Located on California's north coast, the Humboldt Nursery grows
bareroot seedlings of Douglas-fir for spring planting in coastal
and interior regions of Oregon and California. Both survival and
growth of the plantings strongly depend on when the seedlings are
lifted from the nursery beds and put in cold storage. Appropriate
times to lift were determined from the field performances of 33,000
seedlings of 51 seed sources, lifted on different dates and stored
for varying periods. Depending on seed source, the lifting window
ranged from 7 to 18 weeks in early November to late March. The
window tended to widen as source latitude and elevation increased,
but not for all regions. Differences among sources from the same
region were common. Repeated tests of diverse sources have shown
the windows are stable. Seven of 10 sources have windows of 3 to 4
months, enabling the nursery to confine lifting to times when soil
and weather conditions are optimum and to schedule lifting around
sources with narrow windows. Performances for seedlings lifted
outside their windows are mostly poor and often preclude successful
plantation establishment.

6.1 INTRODUCTION

The Humboldt Nursery, one of a number of large Forest Service
nurseries in the Western United States, produces bareroot seedlings
of Douglas-fir (Pseudotsuga menziesii [Mirb.] Franco) for spring
planting on National Forest lands in the Pacific Slope regions of
Oregon and northern California. Its clientele reforests mesic to
xeric sites in a wide array of coastal and interior climates, and
annually requests planting stock for hundreds of seed sources.

Duryea, M.L. and Brown, G.N. (eds.). Seedling physiology and reforestation succes
©1984, Martinus Nijhoff/Dr W. Junk Publishers, Dordrecht/Boston/London. ISBN 978-94-009-6139-5

116

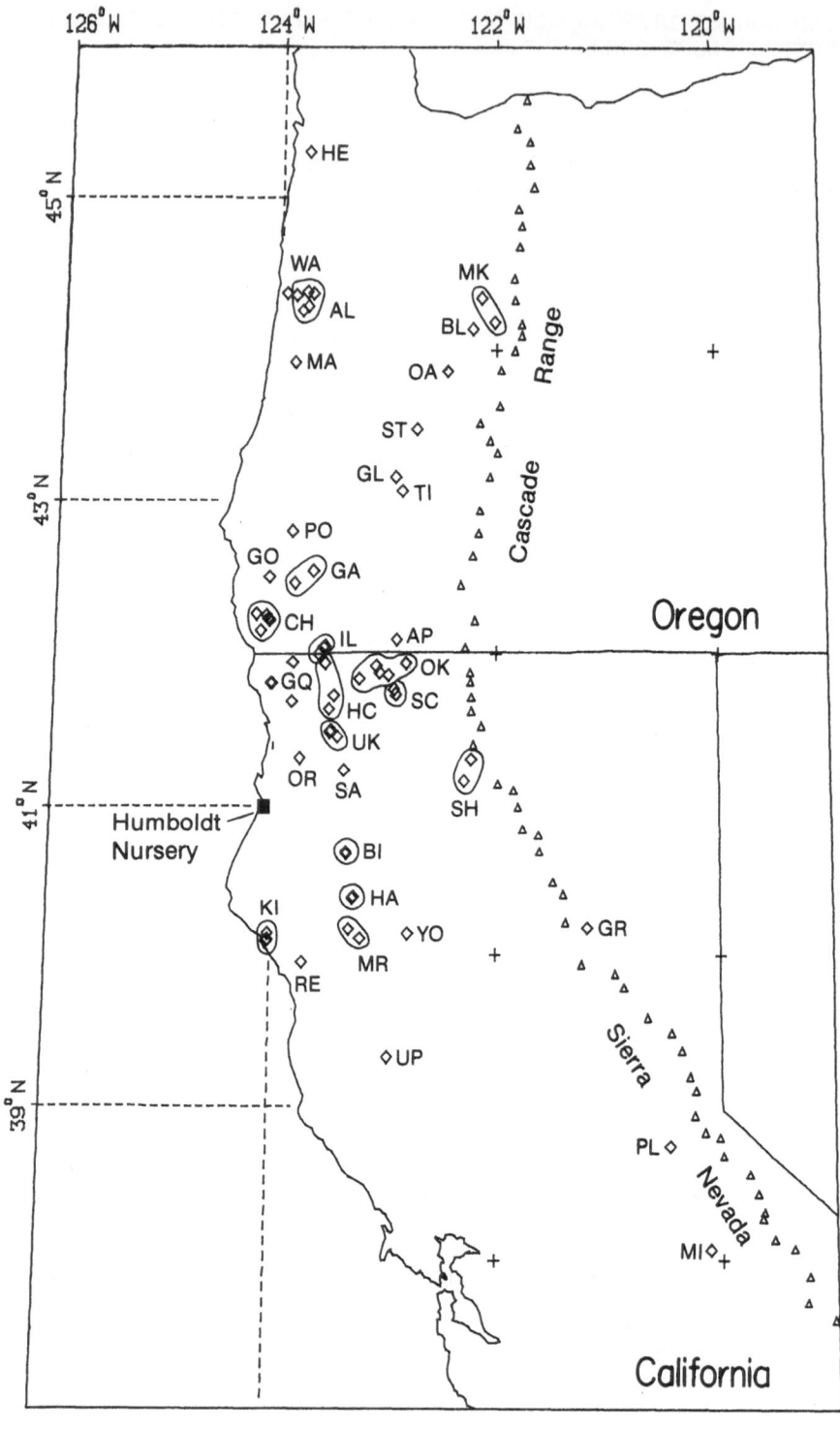

The nursery is located near McKinleyville on California's north coast, and lies on a Pleistocene marine terrace soil 1.6 km (1 mi) inland and 75 m (250 ft) higher than the Pacific Ocean (Fig. 1). Past planting failures with Humboldt stock, especially on sites with cold winters, late springs, and hot, dry summers, were often blamed on the nursery's maritime climate. Mild temperatures in autumn always resulted in a supposed inability to harden seedlings sufficiently prior to winter lifting and cold storage.

Subsequent investigations indicated that field survival could routinely exceed 90 percent in all regions, and showed that the planting success for any particular seed source largely depends on when the seedlings are lifted and stored at the nursery (4, 5). The safe time to lift, that is, the lifting window or sequence of lifting dates for which seedlings survived equally in the field, ranged from less than 2 to more than 4 months in the period from early November to late March.

This paper reports the lifting windows and field performances for coastal and interior sources of Douglas-fir sampled during six winter seasons in the Humboldt Nursery. The trends in latitudinal and elevational variation and the stability of the source lifting windows are summarized. The findings conclusively demonstrate that schedules for winter lifting and cold storage should be keyed to seed source, to ensure the survival and growth potentials necessary for successful reforestation.

6.2 MATERIALS AND METHODS

Lifting windows were determined for a total of 51 seed sources in 66 field tests (Table 1). For each source, seedlings lifted on five or six dates and stored for varying periods were planted on a cleared site in spring and scored for survival through the summer drought. The first test was planted in 1976 and the last in 1981. The windows for 8 sources from different regions were evaluated two or three times, by repeated sowings of the same seed lots.

FIGURE 1. Seed source lifting windows for Douglas-fir in Humboldt Nursery were determined in 66 plantings on the Pacific Slope. (See Table 1 for source and site descriptions.)

Table 1. Seed sources, planting dates, and planting sites used to determine lifting windows for Douglas-fir in the Humboldt Nursery.

Region, district, and seed source [1]			Planting date	Slope (%)	Elev (m) [2]	Lat (°N)	Long (°W)
North Coastal Oregon							
Hebo	HE 053.10	79	May 1	SW,45	245	45.30	123.7(
Waldport	WA 061.10	77	Apr 15	NW,5	275	44.37	123.9!
Alsea	AL 252.15	80	Mar 31	E,30	460	44.37	123.7(
	AL 252.10	77	Apr 21	S,50	245	44.38	123.7(
	AL 252.10	81	Apr 13	N,60	220	44.29	123.7!
	AL 252.05	78	Apr 13	S,30	150	44.36	123.8(
	AL 061.05	79	Apr 10	N,55	150	44.26	123.8(
Mapleton	MA 062.05	79	Apr 24	SW,50	395	43.92	123.8(
South Coastal Oregon							
Powers	PO 072.25	79	Apr 26	NW,30	740	42.80	123.8(
Gold Beach	GO 081.20	79	Apr 5	W,25	550	42.50	124.0(
Galice	GA 511.30	79	Apr 14	W,30	945	42.54	123.6(
	GA 512.25	79	Apr 14	S,20	855	42.46	123.8:
Chetco	CH 082.25	76	Apr 23	W,20	490	42.26	124.1'
	CH 082.25	77	Mar 15	S,30	830	42.22	124.0:
	CH 082.25	78	Apr 6	S,20	700	42.25	124.0(
	CH 082.25	79	Apr 23	S,30	685	42.23	124.0(
	CH 082.10	79	Apr 23	NW,30	335	42.15	124.1.
Coastal California							
Gasquet	GQ 301.30	79	Apr 6	SW,60	760	41.93	123.8.
	GQ 301.30	77	Apr 25	S,15	520	41.81	124.0.
	GQ 301.30	78	May 1	S,15	520	41.81	124.0.
	GQ 091.20	81	Apr 8	S,10	610	41.69	123.8(
Orleans	OR 302.30	79	Apr 4	E,50	915	41.32	123.7(
Mad River	MR 303.45	79	Apr 14	ridge	1220	40.11	123.2(
	MR 340.36	78	Apr 24	ridge	1130	40.17	123.3(
King Range	KI 390.25	77	Mar 18	N,50	610	40.14	124.0:
	KI 390.20	80	Apr 11	ridge	520	40.10	124.0:
	KI 390.20	79	Mar 30	ridge	605	40.09	124.0:
Red Mtn	RE 093.25	78	Apr 6	ridge	550	39.95	123.7:
Upper Lake	UP 372.30	77	Mar 10	ridge	1050	39.32	122.9!

Table 1. Seed sources, planting dates, and planting sites. (cont)

Region, district, and seed source [1]		Planting date	Slope (%)	Elev (m) [2]	Lat (°N)	Long (°W)
Western Klamath Mtns						
Illinois	IL 512.40 79	Apr 24	SE,35	1095	42.05	123.54
	IL 512.35 78	May 16	NW,15	1065	42.04	123.56
	IL 512.13 79	May 8	N,35	610	42.00	123.60
Happy Camp	HC 301.50 79	May 23	bench	1525	41.94	123.54
	HC 301.30 79	Mar 20	ridge	745	41.64	123.50
	HC 301.30 78	Apr 28	E,20	640	41.73	123.46
	HC 301.30 77	Mar 10	E,20	640	41.73	123.46
Ukonom	UK 302.44 79	Mar 24	SW,	1370	41.50	123.48
	UK 311.40 79	Apr 9	SE,35	1220	41.46	123.42
	UK 301.20 79	Mar 23	SE,	610	41.49	123.49
Southern Klamath Mtns						
Salmon River	SA 311.40 79	Mar 25	E,40	1145	41.24	123.36
Big Bar	BI 312.40 77	Mar 17	NW,10	990	40.69	123.33
	BI 312.30 78	May 17	ridge	915	40.68	123.33
Hayfork	HA 312.25 78	Apr 27	ridge	900	40.39	123.26
	HA 312.25 79	Apr 2	ridge	915	40.38	123.27
Yolla Bolla	YO 371.45 78	May 2	N,50	1370	40.14	122.78
Oregon Cascade Range						
McKenzie	MK 472.45 79	Jun 19	SW,35	1280	44.34	122.14
	MK 472.30 80	May 19	NW,60	855	44.18	122.02
Blue River	BL 472.30 77	Apr 8	SW,35	700	44.14	122.22
Oakridge	OA 482.30 81	Mar 27	S,50	795	43.86	122.45
Steamboat	ST 491.30 79	Apr 17	SW,50	730	43.48	122.73
Glide	GL 491.30 79	Jun 5	S,20	945	43.16	122.92
Tiller	TI 492.30 79	Apr 16	SE,20	915	43.07	122.86
Eastern Klamath Mtns						
Applegate	AP 511.40 79	Apr 26	ridge	915	42.09	122.90
Oak Knoll	OK 321.40 77	May 4	S,10	1220	41.95	122.82
	OK 321.40 78	Apr 11	SE,15	1065	41.86	122.97
	OK 321.40 79	Apr 5	S,10	1220	41.92	123.08
	OK 321.30 80	Apr 3	SE,10	1070	41.88	123.05
	OK 321.30 81	Apr 4	NW,15	855	41.84	123.23

120

Table 1. Seed sources, planting dates, and planting sites. (cont)

Region, district, and seed source [1]	Planting date	Slope (%)	Elev (m) [2]	Lat (°N)	Long (°W)
Scott River SC 322.40 78	May 3	ridge	1340	41.77	122.92
SC 322.40 79	May 15	W,30	1220	41.74	122.90
Cascade-Sierra Nevada					
Mt Shasta SH 516.30 77	May 6	bench	1585	41.31	122.22
SH 521.40 79	May 22	bench	1645	41.17	122.28
Greenville GR 523.45 77	Apr 25	SW,10	1310	40.18	121.19
Placerville PL 526.40 77	Apr 1	NE,30	1400	38.75	120.46
Mi-Wok MI 531.40 77	Apr 15	W,30	1525	38.07	120.11

[1] Source codes indicate the district, tree seed zone and elevations in hundred ft (9), and year planted. The districts are listed by latitude in coastal and interior regions. See Figure 1.
[2] 1 m = 3.28 ft.

6.2.1 Seed sources

Seeds of all sources were from operational collections made by the Bureau of Land Management and National Forest Ranger Districts served by the Humboldt Nursery (Table 1, Fig. 1). The seed sources represented one or more tree seed zones (10) in 34 Ranger Districts between latitudes 38° and 46° N, and broadly sampled the elevations of greatest reforestation activity in various climatic regions of the Pacific Coast Ranges, the Klamath Mountains, the Cascade Range, and the Sierra Nevada.

6.2.2 Nursery procedures

Seeds were sown in prepared beds in April or May. The seedlings were mostly grown for 2 years (2-0 stock), and sometimes for 1 year (1-0 stock). The standard cultural regime for the second growing season included vertical root-pruning in spring and one undercut at a depth of 20 cm (8 inches) in summer, timed to improve root growth and control height growth (11).

Contiguous sampling plots were flagged in October. Seedlings were sampled monthly beginning in October or November and ending in

March. For each lift, seedlings were culled to a stem diameter of 4 mm (2-0 stock) or 3 mm (1-0 stock), root-pruned 25 cm (10 inches) below the cotyledon scars, and stored at 1° C (34° F) in standard polyethylene-lined packing bags.

Air temperature in a weather shelter and soil temperature at a depth of 8 cm were recorded continuously from September to April, the period comprising onset and release of seedling dormancy. The natural chilling or cold exposure from October 1 to any particular lifting date was estimated by the cumulative number of hours that air temperature was cooler than 10° C (50° F).

6.2.3 Field procedures

For each source and lifting date, stored seedlings were sorted into 10 replications, packed in a polyethylene-lined bag on ice in a styrofoam cooler, and delivered to the Ranger District of seed origin. District personnel planted the seedlings on cleared sites that had supported Douglas-fir or mixed conifer forest before harvest or wildfire. Planting holes were dug with either hoedads, shovels, or power augers, depending on district. Planting dates ranged from March 10 to June 19, with a median in April (Table 1). For most of the tests, planting was done after the maximum daily soil temperature had reached 5° C (41° F) at a depth of 8 cm, and before the last spring rain.

The planting design consisted of 10 replications of a randomized complete block of lifting date plots, with one row of 10 seedlings per plot. Blocks were either scattered or clustered on the site, depending on the distribution of plantable ground. The seedlings were spaced 0.6 or 1 m (2 or 3 ft) apart, in parallel rows running up the prevailing slope.

About 30 plantings were kept free of invading shrubs, herbs, and grasses the first summer, and 9 were protected against browsing mammals by tubing the seedlings with vexar mesh or poultry wire.

Field survival the first year was determined after autumn rains ended the summer drought. Seedlings in 60 plantings were measured for stem height, diameter, and leader length after the second year. Seedlings in 20 plantings were remeasured after 3 and 4 years.

122

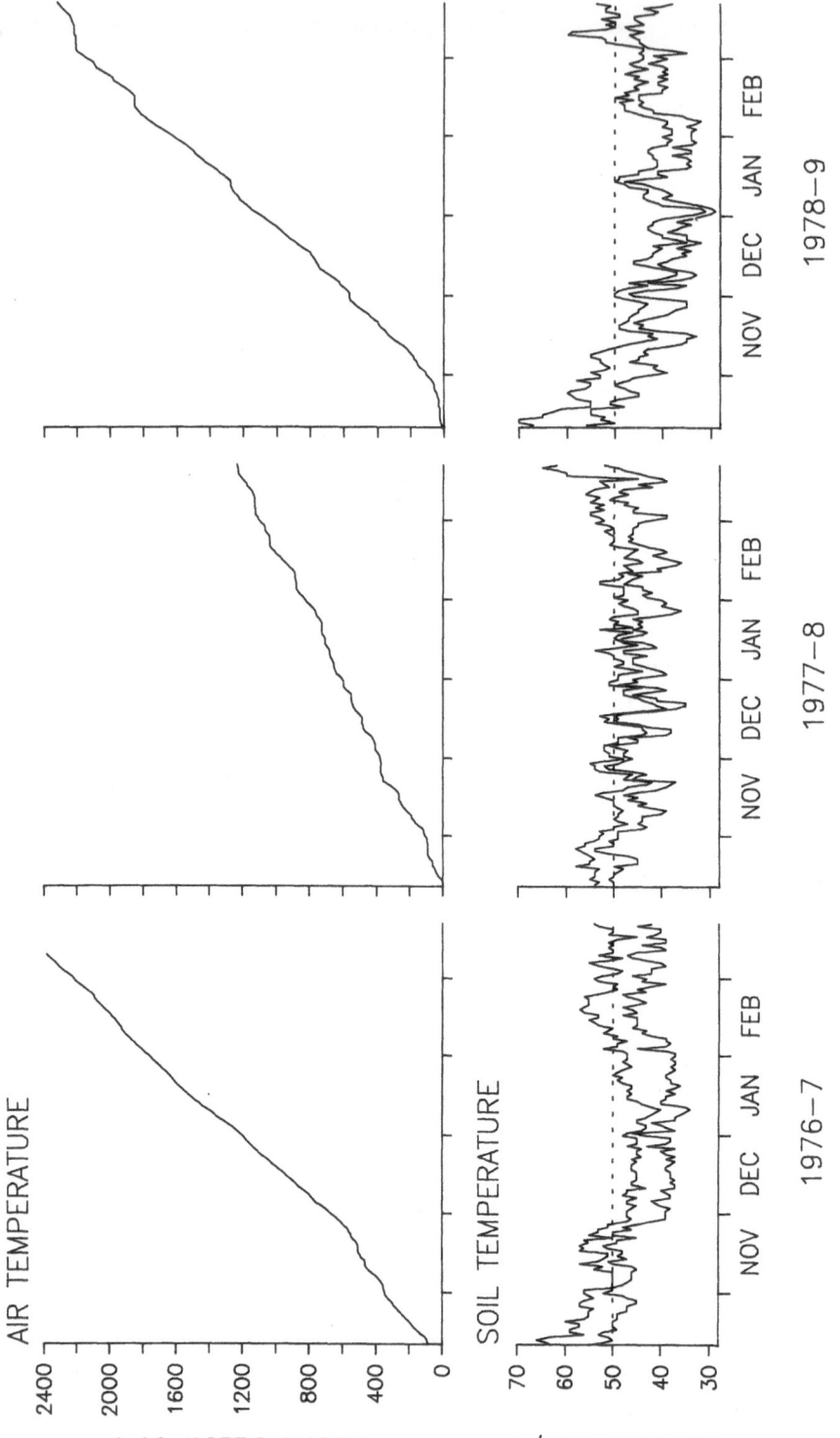

FIGURE 2. Seedlings in the Humboldt Nursery were exposed to colder temperatures in the winter seasons of 1976-7 and 1978-9 than in 1977-8, when the winter was abnormally warm.

6.2.4 Analyses

The effects of lifting date on survival and growth were assessed by analyses of variance (7). For each source and trait, the least significant difference at the 5 percent level was calculated from the error mean square, with LSD = $q[ems/r]^{1/2}$, r = 10 observations per mean, and q = 2.87 for 36 degrees of freedom (9).

The source lifting window was defined as the interval between the first and last safe lift dates. The safe dates were estimated by graphing field survival against lifting date and determining the value of X for Y = highest survival - LSD. Associations between width of the window and source elevation, latitude, and longitude were measured by multiple correlation coefficients (R).

6.3 RESULTS

Most seedlings stopped flushing and set buds in October. Root elongation ceased sometime in November and mostly resumed in late March. With winter soil temperatures below 10° C (50° F), visible traces of root growth consisted of white root tips shorter than 1 or 2 mm.

The winter lifting season was much cooler in 1976-7 and 1978-9 than in 1977-8 (Fig. 2). Natural chilling totalled 600 hours in the cool autumns (to December), as against 400 hours in the warm autumn. The minimum daily soil temperature commonly dropped below 10° C in late October and normally remained low until April. The maximum daily soil temperature dropped below 10° C in mid- to late November, often was above 10° C in the warm but not cool winters, and exceeded 10° C in late February or early March.

6.3.1 Window variability

Field survival showed that the safe times to lift seedlings for cold storage and spring planting depended markedly on seed source (Table 2, Fig. 3). Survival was zero for all lifts made before the middle of October, but increased at various rates for subsequent lifts. Source lifting windows opened on dates ranging from early November to late January and closed sometime in February to late March. The widest window for any source was estimated at 127 days (OK 321.40) and the narrowest at 49 days (CH 082.25).

124

Table 2. Seed source lifting windows for Douglas-fir in the Humboldt Nursery.[1]

Field survival (percent by nursery lifting date)[2]

Region and seed source	November 1	November 30	December 31	January 31	February 28	March	LSD[3]
North Coastal Oregon							
HE 053.10 79	79.	99.	99.	99.	97.	97.	7.9
WA 061.10 77	35.		89.	94.	86.	96.	9.8
AL 252.15 80[4]		72.	99.	99.	100.		9.8
AL 252.10 77	60.		97.	97.	98.	100.	8.2
AL 252.10 81	57.		95.	98.	96.	98.	5.6
AL 252.05 78		86.	96.	99.	100.	100.	7.5
AL 061.05 79	82.		100.	98.	98.	98.	6.8
MA 062.05 79	77.		87.	97.	93.	93.	9.1
South Coastal Oregon							
PO 072.25 79	42.		90.	97.	96.	96.	9.5
GO 081.20 79	29.		64.	72.	67.	60.	15.4
GA 511.30 79		73.	82.	84.	81.	92.	13.8
GA 512.25 79		74.	91.	92.	90.	88.	9.8
CH 082.25 76	0.		28.	73.	96.	84.	10.2
CH 082.25 77	13.		55.	69.	82.	76.	12.6
CH 082.25 78		70.	64.	67.	88.	89.	14.4
CH 082.25 79	52.		83.	93.	89.	93.	9.5
CH 082.10 79	42.		87.	95.	91.	83.	9.5
Coastal California							
GQ 301.30 79[5]	7.	25.	75.	64.	48.		19.3
GQ 301.30 77	88.	98.	98.	97.	98.		4.6
(row cut off at page edge)		85.	98.		99.	97.	7.8

Site							
GQ 091.20 81	70.		94.	92.	97.	97.	6.0
OR 302.30 79	67.		73.	81.	78.	89.	11.4
MR 340.36 78	64.		74.	88.	92.	91.	13.0
MR 303.45 79	40.		55.	74.	75.	66.	13.6
KI 390.25 77	5.		45.	67.	66.	81.	16.2
KI 390.20 80	4.		81.	93.	94.	94.	9.5
KI 390.20 79	79.		88.	85.	90.	87.	10.2
RE 093.25 78	20.		69.	82.	75.	86.	16.9
UP 372.30 77	6.		45.	56.	48.	51.	11.2
Western Klamath Mtns							
IL 512.40 79	34.	85.	73.	61.	71.	67.	17.7
IL 512.35 78	16.		59.	55.	61.	43.	16.3
IL 512.13 79	88.		28.	58.	56.	51.	13.4
HC 301.50 79			97.	98.	97.	100.	5.1
HC 301.30 79	59.	89.	93.	91.	89.	91.	11.2
HC 301.30 78	72.		87.	95.	94.	93.	10.7
HC 301.30 77	38.		95.	98.	97.	96.	10.5
UK 302.44 79	86.						6.0
UK 311.40 79	30.		90.	90.	71.	74.	16.3
UK 301.20 79	54.		80.	90.	93.	93.	11.5
Southern Klamath Mtns							
SA 311.40 79	87.	78.	97.	96.	99.	100.	6.4
BI 312.40 77	45.		85.	98.	94.	96.	13.3
BI 312.30 78	87.	78.	87.	93.	86.	90.	8.0
HA 312.25 78	73.		87.	93.	85.	94.	9.8
HA 312.25 79	64.		96.	93.	94.	95.	9.3
HA 312.25 79	73.						10.0
YO 371.45 78	67.	90.	90.	84.	93.	95.	9.9

Oregon Cascade Range

Plot						
MK 472.45 79	75.	79.	84.	75.	84.	10.5
MK 472.30 80 [4]	48.	88.	86.	88.	88.	14.9
BL 472.30 77	64.	90.	80.	88.	76.	12.9
OA 482.30 81	37.	74.	69.	77.	80.	9.5
ST 491.30 79	87.	85.	90.	88.	92.	11.7
GL 491.30 79	60.	85.	92.	89.	87.	13.0
TI 492.30 79	97.	97.	98.	93.	100.	5.4

Eastern Klamath Mtns

Plot						
AP 511.40 79	69.	86.	90.	93.	90.	11.6
OK 321.40 77 [5]	1.	39.	55.	71.	52.	19.0
OK 321.40 78	92.	88.	88.	95.	90.	7.4
OK 321.40 79 [4]	83.	98.	97.	97.	97.	8.3
OK 321.30 80 [4]	58.	73.	88.	84.	79.	13.9
OK 321.30 81	54.	87.	78.	83.	83.	10.1
SC 322.40 78 [5]	44.	92.	90.	84.	92.	8.9
SC 322.40 79	19.	64.	84.	81.	63.	13.5

Cascade-Sierra Nevada

Plot						
SH 516.30 77 [4]	7.	41.	71.	71.	71.	14.1
SH 521.40 79 [4]	78.	89.	88.	90.	91.	11.6
GR 523.45 77	1.	39.	66.	84.	73.	11.5
PL 526.40 77	0.	77.	92.	90.	74.	12.8
MI 531.40 77	70.	87.	90.	93.	73.	12.0

1 Seedlings stored at 1° C (34° F) were planted in spring. See Table 1.
2 Bars show the lifting windows. Points are dates that seedlings were sampled (n = 100).
3 Least significant difference at the 5 percent level.
4 Lifted after one growing season.
5 Planted on landslides (GQ) and ultramafic soils (OK, SC).

Mean survival for seedlings lifted within their window was 80 to 99 percent in 48 tests, 60 to 80 percent in 15 others, and 50 to 55 percent in 3 tests. Low survival was commonly related to problems other than lifting date. Severe browsing and competing vegetation were frequent causes of mortality, but improper planting time (too early or late), poor root placement, and various offsite plantings were also encountered.

Width of the lifting window was not correlated with survival. High survival was obtained for sources with wide windows and for sources with narrow windows. In the 1978 plantings, for example, sources IL 512.35 and HA 312.25 had windows of 101 and 106 days, but survivals of 55 and 89 percent; CH 082.25 and OK 321.40 had windows of 49 and 120 days, with survivals of 88 and 90 percent. In 1979, IL 512.40 and SA 311.40 had windows of 117 and 116 days, but survivals of 71 and 97 percent; CH 082.10 and HC 301.30 had windows of 83 and 114 days, with survivals of 91 and 92 percent.

6.3.2 Window stability

For plantings on typical Douglas-fir sites, the lifting windows were stable (Table 3, Fig. 3). The first safe date, window width, and even survival were consistent from year to year. For the eight sources in which the same seed lot was used in different years, the difference in the first safe date ranged from 1 day (AL 252.10) to 17 days (CH 082.25). Variations were 1 to 12 percent for the wide windows, and 12 to 35 percent for the narrow window.

For offsite plantings on landslides and ultramafic soils, the lifting windows of three sources (GQ 301.30, OK 321.40, SC 322.40) were narrowed from about 4 months to 2 months in late December to late February (Table 2).

6.3.3 Natural chilling before cold storage

The amount of cold exposure that preceded the lifting windows ranged from 225 hours to 1480 hours (Table 3). The minimum was less than 400 hours in 26 percent of all tests, more than 800 hours in 21 percent, and between 400 and 800 hours in the remainder.

128

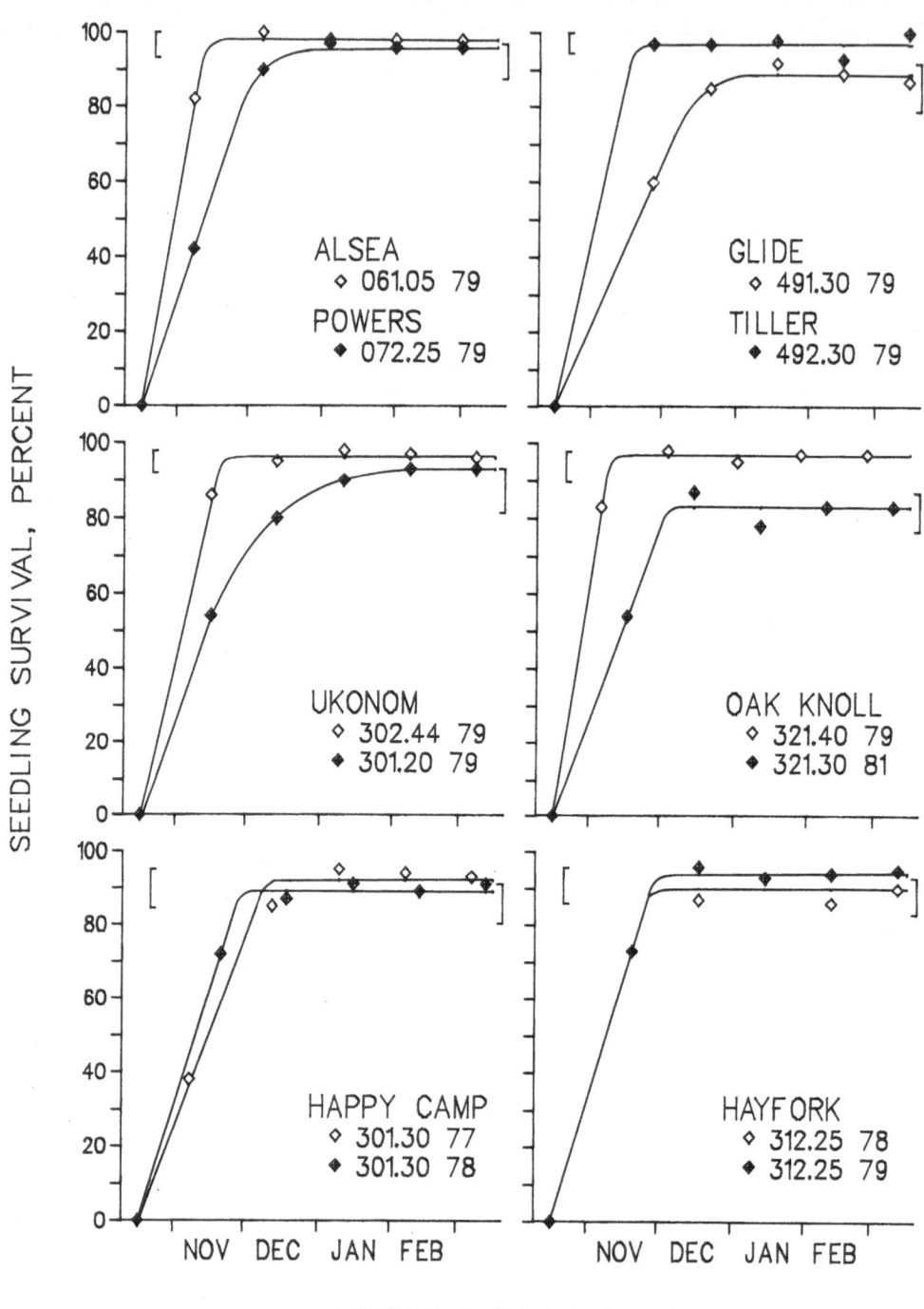

SEEDLING SURVIVAL, PERCENT

ALSEA
⬦ 061.05 79
POWERS
◆ 072.25 79

GLIDE
⬦ 491.30 79
TILLER
◆ 492.30 79

UKONOM
⬦ 302.44 79
◆ 301.20 79

OAK KNOLL
⬦ 321.40 79
◆ 321.30 81

HAPPY CAMP
⬦ 301.30 77
◆ 301.30 78

HAYFORK
⬦ 312.25 78
◆ 312.25 79

NOV DEC JAN FEB

LIFTING DATE IN NURSERY

Table 3. Stability of the lifting window for diverse seed sources
of Douglas-fir in the Humboldt Nursery.

Region and seed source [1]	Window width (days)	Seedling survival (percent)	First safe lift date	Natural chilling [2] (hours)
North Coastal Oregon				
AL 252.10 77	112	98	Nov 24	590
81	111	97	Nov 25	---
South Coastal Oregon				
CH 082.25 76	55	89	Jan 17	---
77	66	76	Jan 11	1481
78	49	88	Jan 26	797
Coastal California				
GQ 301.30 77	124	97	Nov 12	491
78	111	98	Nov 25	376
Western Klamath Mtns				
HC 301.30 77	102	92	Dec 4	777
78	110	89	Nov 26	376
79	114	92	Nov 22	459
Southern Klamath Mtns				
HA 312.25 78	106	89	Nov 30	387
79 [3]	110	90	Nov 26	542
79	110	94	Nov 26	542
Eastern Klamath Mtns				
OK 321.40 78	120	90	Nov 16	277
79	127	96	Nov 9	225
OK 321.30 80 [3]	96	81	Dec 10	---
81	105	83	Dec 1	---

[1] See Table 1 and Figure 1.
[2] Cumulative number of hours that air temperature was colder than
10° C (50° F), from Oct 1 to first safe lift date.
[3] Lifted after one growing season.

FIGURE 3. Seed source and lifting date affected 1-year survival
for Douglas-fir from the Humboldt Nursery. Examples show the range
of source lifting windows for the Oregon Coast and Cascade Ranges
(top panels), wider lifting windows for higher source elevation in
the western and eastern Klamath Mountains (middle panels), and the
stability of lifting windows for the western and southern Klamath
Mountains (bottom panels). The brackets show LSD at the 5 percent
level. (See Tables 1 and 2.)

130

6.3.4 Latitudinal and elevational trends

The width of the lifting window varied among and within regions
About 14 percent of the variation was explained by source latitude
longitude, and elevation, and 13 percent by latitude and elevation
only (R^2 significant at the 5 percent level). For 13 districts
that planted sources from two different elevations, 40 percent of
the variation was explained by source latitude and elevation, and
mostly by elevation (R^2 significant at the 1 percent level).

The window tended to increase in width with source latitude in
the Pacific Coast and Sierra Nevada-Cascade regions, but not in th
southern and western Klamath Mountains (Table 2). Northward in th
coastal regions (Fig. 1), the average window increased from 85 day
for Upper Lake to Orleans (UP, OR) to 113 days for Mapleton to Heb
(MA, HE). In the interior regions, the average increased from 83
days for Mi-Wok to Mt Shasta (MI, SH) to 108 days for Tiller to
McKenzie (TI, MK). Between these coastal and interior transects,
and opposing the overall trend, the windows averaged 114 days for
Yolla Bolla to Salmon River (YO, SA) and 103 days for Ukonom to
Illinois Valley (UK, IL).

For interior districts, the window increased by 12 to 39 days
for increases of 300 to 600 m (1000 to 2000 ft) in source elevatio
(Table 2, Fig. 3). For coastal districts, the window tended eithe
to decrease or remain the same with higher source elevation. The
windows for sources from similar elevations on adjacent districts
differed by 1 to 25 days (Table 2).

6.3.5 Plantation growth

Browsing deer, elk, and cattle destroyed individual plantings o
five separate districts, from Waldport to Placerville (WA to PL).
Deer periodically ate the new leaders and lateral shoots in others
reducing growth and survival in every region. Partly because of
the differential damage, the mean tree height for seedlings lifted
within their source windows ranged from 0.4 to 1.1 m after 3 years
and from 0.5 to 2.2 m after 4 years (Table 4). Leader growth the
fourth year increased tree heights by 45 to 70 percent, depending
on plantation.

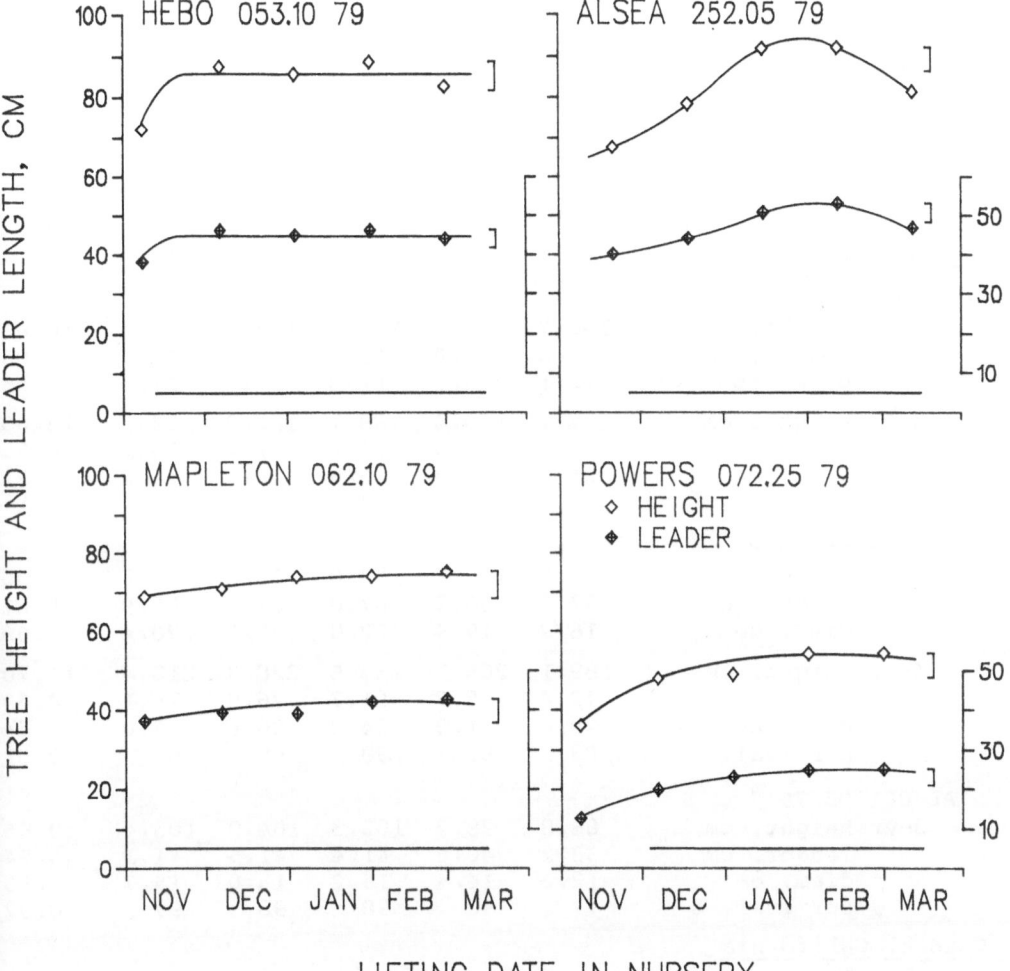

FIGURE 4. Seed source and lifting date affected the 2-year growth of Douglas-fir from the Humboldt Nursery. Examples for sources in the Oregon Coast Range show patterns typical of all regions (left and bottom panels) and one unique pattern (top right panel). The Brackets show LSD at the 5 percent level. Bars show the source lifting windows. (See Tables 1 and 2.)

Seedlings lifted within their source window often grew more in both height and diameter than seedlings lifted outside their window (Fig. 4). Lifting date did not affect growth within the window, except for one source in the Oregon Coast Range. Tree heights in that planting differed among lifting dates by up to 18 percent at 2 years and 14 percent at 4 years (Table 4).

Table 4. Typical effects of seed source and lifting date on 3- and 4-year field performance of Douglas-fir from the Humboldt Nursery.

Region, seed source,[1] and trait	Nov	Dec	Jan	Feb	Mar	LSD[2]
North Coastal Oregon						
AL 252.10 77 [3]						
3-yr height, cm	106.7	105.8	122.6	113.5	111.7	10.66
leader, cm	43.8	42.6	49.1	42.7	45.4	4.63
diam, mm	15.1	15.3	17.8	17.4	17.1	1.58
4-yr height, cm	177.1	170.4	188.7	182.0	179.6	14.00
leader, cm	72.6	67.5	72.1	71.2	70.1	6.29
diam, mm	21.9	22.4	26.6	25.1	24.2	2.27
survival, %	56	89	89	88	93	10.1
AL 252.05 78 [3]						
3-yr height, cm	120.2	133.6	153.6	153.2	139.8	9.36
leader, cm	57.3	60.2	67.0	67.2	62.9	5.62
diam, mm	18.2	19.4	22.0	22.3	20.4	1.60
4-yr height, cm	189.1	206.1	234.6	228.3	210.4	11.70
leader, cm	72.4	75.3	81.7	76.9	75.1	4.63
diam, mm	29.4	31.9	34.3	36.6	35.0	3.44
survival, %	83	95	99	99	100	9.31
AL 061.05 79 [3]						
3-yr height, cm	88.8	98.2	103.3	104.0	103.4	10.46
leader, cm	38.2	40.8	41.4	41.5	41.1	5.94
diam, mm	12.2	14.4	15.2	15.0	15.7	1.55
survival, %	77	98	98	98	97	8.32
Coastal California						
KI 390.25 77						
3-yr height, cm		57.3	55.1	48.0	50.0	6.87
leader, cm		10.7	11.3	11.5	12.0	3.34
diam, mm		13.6	13.3	12.8	13.6	2.20
4-yr height, cm		71.9	67.3	61.1	64.8	8.86
leader, cm		21.2	18.6	20.7	21.3	4.65
diam, mm		19.2	17.8	18.2	17.7	2.47
survival, %	2	44	57	62	62	17.1
Western Klamath Mtns						
HC 301.30 77						
3-yr height, cm	46.4	51.6	51.9	49.4	48.8	4.89
leader, cm	17.5	18.2	21.4	18.5	19.6	2.50
diam, mm	13.0	14.3	14.9	13.6	13.2	1.13
4-yr height, cm	76.8	82.9	88.5	82.6	83.0	8.66
leader, cm	29.2	31.8	36.8	34.3	34.7	3.90
diam, mm	19.2	21.5	22.3	20.9	20.6	1.84
survival, %	38	80	92	93	94	10.9

Table 4. (cont)

Region, seed source,[1] and trait	Nursery lifting date					LSD [2]
	Nov	Dec	Jan	Feb	Mar	
HC 301.30 78						
3-yr height, cm	48.6	51.5	55.2	54.4	55.3	5.77
leader, cm	18.6	19.6	21.8	19.7	21.2	2.86
diam, mm	10.8	11.5	12.5	12.0	12.4	1.26
survival, %	63	84	90	86	90	11.4
Eastern Klamath Mtns						
OK 321.40 78						
3-yr height, cm	40.4	40.4	42.6	39.6	40.7	3.99
leader, cm	19.1	18.0	19.0	17.5	19.4	1.98
diam, mm	10.9	11.3	12.0	11.6	11.5	0.95
survival, %	90	88	87	93	87	8.85
Cascade-Sierra Nevada						
GR 523.45 77 [4]						
3-yr height, cm		36.6	33.8	37.8	40.4	6.32
leader, cm		10.4	12.4	12.7	12.8	3.66
diam, mm		12.9	11.4	13.0	13.2	1.79
4-yr height, cm		52.0	48.2	53.1	57.9	10.80
leader, cm		13.2	13.9	16.6	18.5	6.50
diam, mm		16.3	15.2	16.6	17.3	3.03
survival, %	0	26	54	65	64	13.7

[1] See Tables 1 and 2.
[2] Least significant difference at the 5 percent level.
[3] Screened against browsing mammals.
[4] Screened after severe browsing in second spring.

The effects of lifting date on seedling growth necessarily were assessed in the plantings that had minimal or no browse damage. In those plantings, leader growth the second year increased seedling heights by 27 to 74 percent for sources in the Klamath Mountains of California, and by 57 to 125 percent for sources in the Coast and Cascade Ranges of Oregon (Table 5).

Table 5. Typical effects of seed source and lifting date on the
2-year field performance of Douglas-fir from the Humboldt Nursery.

Region, seed source, [1] and trait	Nursery lifting date					LSD [2]
	Nov	Dec	Jan	Feb	Mar	
North Coastal Oregon						
HE 053.10 79 [3]						
2-yr height, cm	72.1	87.5	85.7	88.6	82.7	7.08
leader, cm	38.2	46.3	45.0	46.3	44.1	4.61
diam, mm	11.0	13.7	14.1	14.4	13.4	1.29
survival, %	77	99	99	97	97	7.76
MA 062.10 79 [3]						
2-yr height, cm	68.8	70.9	73.9	74.0	75.2	7.03
leader, cm	37.3	39.4	39.1	42.0	42.7	5.88
diam, mm	14.8	14.9	15.1	15.8	16.5	1.76
survival, %	77	83	94	91	93	9.62
South Coastal Oregon						
PO 072.25 79 [3]						
2-yr height, cm	36.1	48.0	49.0	54.2	54.3	6.20
leader,cm	12.8	19.9	23.1	24.9	25.1	4.23
diam, mm	6.6	9.0	9.8	11.3	10.9	1.32
survival, %	41	88	96	93	94	10.3
CH 082.25 79						
2-yr height, cm	44.1	49.4	51.8	43.8	50.4	7.24
leader, cm	18.8	18.2	20.2	15.0	18.3	5.20
diam, mm	9.7	11.4	11.6	9.5	10.8	1.54
survival, %	50	81	93	87	92	10.6
Western Klamath Mtns						
HC 301.50 79						
2-yr height, cm	29.0	32.7	30.9	30.0	31.6	3.06
leader, cm	8.6	9.1	8.6	8.7	8.6	1.37
diam, mm	7.3	8.6	7.7	7.9	8.1	0.81
survival, %	80	94	80	82	94	10.8
HC 301.30 79						
2-yr height, cm	54.5	59.9	55.5	57.6	55.2	4.71
leader, cm	14.3	15.7	13.0	12.8	12.3	3.42
diam, mm	11.9	13.6	12.1	13.9	11.7	1.52
survival, %	58	87	91	96	88	11.0
UK 302.44 79						
2-yr height, cm	34.0	32.6	34.9	33.7	32.5	4.14
leader	13.6	10.7	11.2	11.7	10.4	3.28
diam, mm	9.7	9.4	9.9	9.9	9.5	1.14
survival, %	80	90	93	91	92	8.57

Table 5. (cont)

Region, seed source, [1] and trait	Nursery lifting date					LSD [2]
	Nov	Dec	Jan	Feb	Mar	
UK 301.20 79						
2-yr height, cm	46.2	51.5	52.9	53.5	47.7	6.30
leader, cm	9.8	11.3	11.5	11.3	9.3	4.19
diam, mm	8.9	9.7	10.4	10.0	9.7	1.13
survival, %	52	78	87	88	89	11.6
Southern Klamath Mtns						
SA 311.40 79						
2-yr height, cm	43.2	44.5	44.4	44.5	46.5	4.34
leader, cm	8.8	10.1	10.0	10.0	11.8	1.87
diam, mm	9.2	9.9	9.9	10.6	11.0	0.94
survival, %	82	93	93	98	98	8.36
Oregon Cascade Range						
MK 472.45 79						
2-yr height, cm	34.6	37.1	38.1	37.4	37.1	3.65
leader, cm	12.6	13.2	14.8	14.9	12.7	2.67
diam, mm	9.4	10.7	10.8	11.6	10.5	1.44
survival, %	60	64	76	66	67	10.8
Eastern Klamath Mtns						
OK 321.40 79						
2-yr height, cm	39.6	43.6	44.6	46.8	44.0	4.53
leader, cm	16.6	18.8	18.3	19.9	19.3	2.45
diam, mm	7.4	8.4	8.5	8.4	7.9	1.04
survival, %	81	97	95	97	96	8.27

[1] See Tables 1 and 2.
[2] Least significant difference at the 5 percent level.
[3] Screened against browsing mammals.

6.4 DISCUSSION

All seed source lifting windows determined for Douglas-fir in the Humboldt Nursery overlapped for a period of 4-5 weeks in late January and early February. Field survivals in every test on the Pacific Slope confirmed the generalization that seedlings stored in late winter will be characterized by high survival potentials (1).

136

6.4.1 <u>Source windows and lifting schedules</u>

Any nursery that produces seedlings in the millions must expect,
and surely dreads, the yearly avalanche of clientele requests to
lift all stock in late winter only. Every forester rightly demands
dormant seedlings that have good balance between pruned roots and
transpiring tops, large root systems with abundant root tips, and
high capacities for root and top growth.

To ensure root function and survival potential, seedlings must
be lifted with little or no damage to the roots that remain above
pruning depth. At the Humboldt Nursery, winter rains are common
and make the beds soggy for days at a time. Lifting seedlings when
the soil is wet, heavy and sticky, will unavoidably strip secondary
roots and break conductive tissues of the primary laterals. Such
losses are intolerable because planted seedlings avoid prolonged
water stress and survive the summer drought largely by elongation
from intact root tips.

By utilizing seed source lifting windows, Humboldt Nursery can
schedule safe liftings from late November to late March, and plan
to lift only when the soil conditions make root damage unlikely.
Sources with narrow windows are the critical ones, because all of
their seedlings must be lifted in late winter to have high survival
potentials at planting time in spring. Sources with wide windows
are most flexible, because dormant seedlings may be lifted anytime
during the winter season and still have high survival potentials
after cold storage. More than one-fourth of the sources evaluated
for Humboldt Nursery had windows open by November 22, and more than
two-thirds had windows open by December 7 (Table 2).

To simplify planning a lifting schedule, sources can be grouped
into window types (Table 6): Type 1 includes every source with a
window that opens by November 21. Type 2 includes all windows that
open in the 2 weeks after November 21; type 3, in the 2 weeks after
December 6; type 4, in the 2 weeks after December 21; and type 5,
after January 5. Seedlings of type 1 sources can be confidently
stored beginning December 1, of type 2, December 10, and so on.
Seedlings of every source should be stored before late March, and
sooner for certain sources of types 3 to 5, because all seedlings
deharden with the onset of spring and quickly resume root growth.

Table 6. Types of seed source lifting windows for Douglas-fir in the Humboldt Nursery.

Lifting window type	Store seedlings after	Sources included [1] (percent)	First safe lift date [2]	Window width (days)
1	Nov 30	28	Nov 7-21	127 to 115
2	Dec 10	40	Nov 22-Dec 6	114 to 100
3	Dec 25	20	Dec 7-21	99 or less
4	Jan 10	8	Dec 22-Jan 5	85 or less
5	Feb 1	4	Jan 6-26	70 or less

[1] n = 51.
[2] Last safe date is Mar 16, with exceptions. See Table 2.

6.4.2 Natural chilling and cold storage

Seedling dormancy in the Humboldt Nursery is induced by either moderate summer water stress, the autumn decrease in photoperiod, or modest cold exposure. For most sources of Douglas-fir, the chilling received before the lifting window opened was clearly minimal. One-fourth of the sources evaluated were safely stored with less than 400 hours of chilling. Three-fourths were stored with less than 800 hours, which included the first safe dates for all type 1 and 2 windows and half of the type 3 windows (Table 6). In the warmest lifting season, chilling did not total 800 hours until February (Fig. 2, 1977-8), after the narrowest window had already opened (Table 3, CH 082.25).

Field survival and growth have consistently demonstrated the efficacy of overwinter cold storage for Douglas-fir. At planting time in spring (Table 1), seedlings lifted earliest within their source window (Table 2) had been in storage for periods ranging from 7 weeks (CH 082.25) to 7 months (MK 472.45). Storage at 1° C has always completed the chilling that seedlings from early lifts need to release dormancy and promote shoot growth (4, 5).

Seedlings were stored safely for more than 3 months in 56 tests, 4 months in 31 tests (for example, Fig. 3 and 4), and 5 months in 5 tests. Had seedlings been lifted on the first safe date, the time

138

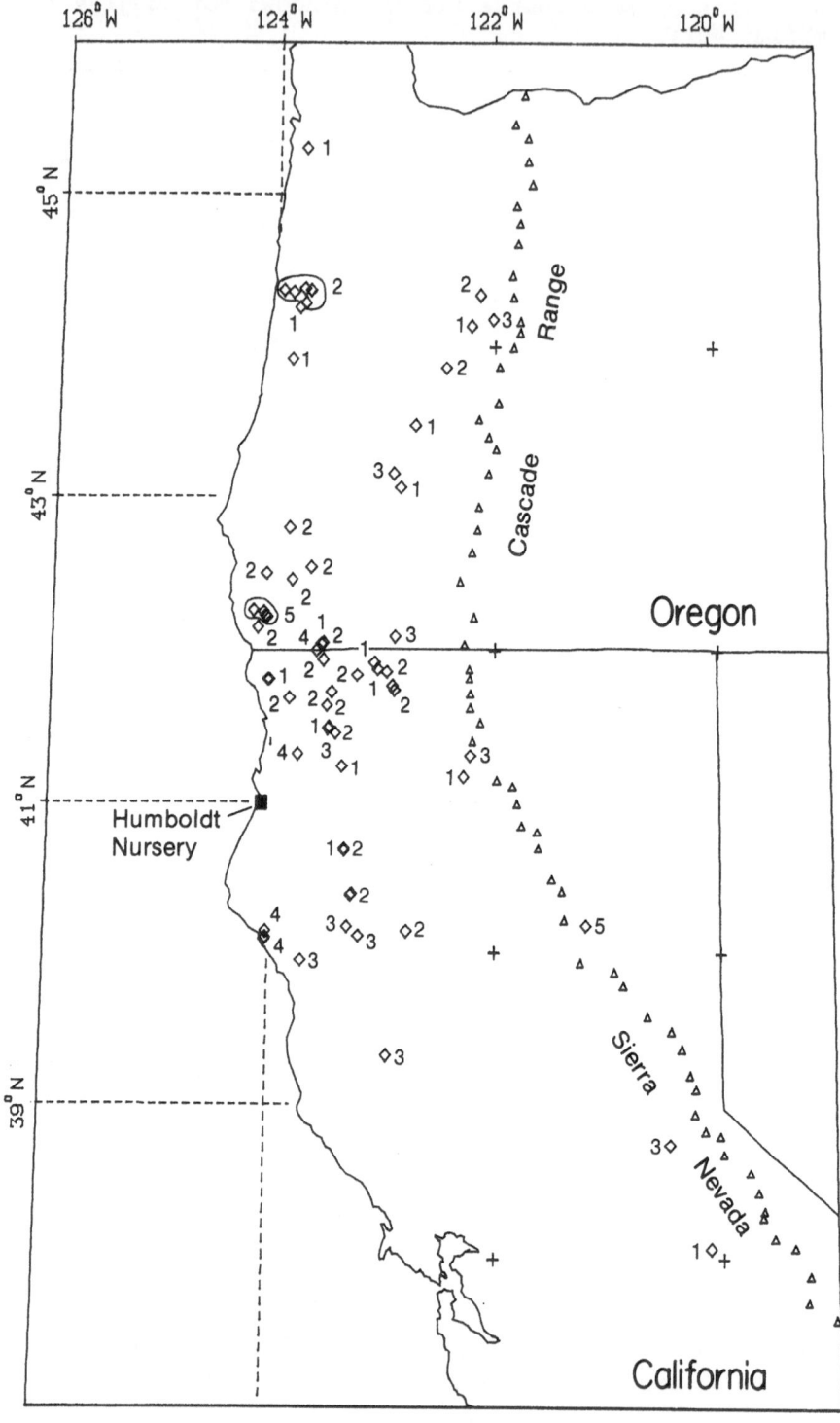

in storage would have exceeded 4 months in 49 tests and 5 months in 17 tests. Any longer period of storage would rarely be necessary, because tests were normally installed at the close of the planting season, after local cooperators had finished their spring planting programs.

6.4.3 Scheduling untested sources

Safe times to lift the many untested sources that the Humboldt Nursery grows should be guided by the windows of known sources from the same region (Fig. 5). Untested sources from the Oregon coastal regions may be confidently lifted as window type 2, because the window of every known source is type 1 or 2, except for one type 5 (CH 082.25). Untested sources from the California coastal regions should initially be lifted as type 3, except for those from areas of known type 4. New sources from the Klamath Mountain regions can be safely lifted as type 2, except for those of lowest elevations, peripheral districts, and marginal sites, which should be treated as type 3. New sources from the Cascade regions should initially be lifted as type 3, which seems as common as either type 1 or 2. Untested sources from the Sierra Nevada must currently assume the type 5 window at Humboldt Nursery.

6.5 CONCLUSIONS

Every forest tree nursery that produces bareroot planting stock for a variety of sites and climates should determine their own seed source lifting windows, to ensure appropriate lifting schedules and to improve the probability of successful plantation establishment. Results for Douglas-fir in the Humboldt Nursery and other research on cold storage of bareroot seedlings show that knowledge of their lifting windows could help each Forest Service nursery and private regional supplier in the Western United States.

FIGURE 5. The Humboldt Nursery ensures high survival potentials for Douglas-fir planting stock by lifting sources of window type 1 after Nov 30; 2, after Dec 10; 3, Dec 25; 4, Jan 10; and 5, Feb 1.

140

Tests for the Albuquerque Nursery, which produces ponderosa pin
(Pinus ponderosa Dougl. ex Laws.) for reforestation in New Mexico
and Arizona, indicate a wide variation in source lifting windows
(3). Similar tests for the Wind River Nursery in Washington and
Lucky Peak Nursery in Idaho, which contend with frozen soils and
winter snows, might beneficially expand their fall lifting seasons
for Douglas-fir and ponderosa pine by identifying all sources that
store safely in late autumn or early winter (2, 8). Tests for J.
Herbert Stone Nursery in southern Oregon and for the Placerville
Nursery in the western Sierra Nevada of California have repeatedly
shown high survival for various sources of ponderosa pine stored i
late November (6).

REFERENCES

1. Cleary, B. D., R. D. Greaves, and P. W. Owston. 1978.
    Seedlings. In Regenerating Oregon's forests, B. D.
    Cleary, R. D. Greaves, and R. K. Hermann, eds. p. 63-97.
    Oregon State Univ. Ext. Serv., Corvallis.
2. Deffenbacher, F. W., and E. Wright. 1954. Refrigerated
    storage of conifer seedlings in the Pacific Northwest.
    J. Forestry 52: 936-938.
3. Heidmann, L. J. 1983. An initial determination of the
    "lifting window" for ponderosa pine seedlings raised at
    Albuquerque, New Mexico. In Proc. Western Nurserymen's
    Conf., Medford, Oregon, August 10-12, 1982. p. 77-84.
    Southern Oregon Regional Services Institute, Southern
    Oregon State College, Ashland.
4. Jenkinson, J. L., and J. A. Nelson. 1978. Seed source
    lifting windows for Douglas-fir in the Humboldt Nursery.
    In Western Forest Nursery Council and Intermountain
    Nurseryman's Assoc. combined Nurseryman's Conf. and Seed
    Processing Workshop, Eureka, Calif., August 7-11, 1978.
    p. B77-95. USDA Forest Serv., Region 5, San Francisco.
5. Jenkinson, J. L., and J. A. Nelson. 1983. 1-0 Douglas-
    fir: a bare-root planting option. In Proc. Western
    Nurserymen's Conf., Medford, Oregon, August 10-12, 1982.
    p. 63-76. Southern Oregon Regional Services Institute,
    Southern Oregon State College, Ashland.
6. Jenkinson, J. L., and M. J. Knight. 1984. Seed source
    and nursery affect growth capacity and field survival of
    1-0 ponderosa pine. Manuscript submitted to Forest
    Science.
7. Jennrich, R., and P. Sampson. 1981. P8V. In BMDP
    statistical software, W. J. Dixon and others, eds.
    p. 427-435. Univ. California Press, Berkeley.
8. Morby, F. E., and R. A. Ryker. 1979. Fall-lifted
    conifers successfully spring planted in southwest Idaho.
    Tree Planters' Notes 30(3): 27-29.

9. Steel, R. G. D., and J. H. Torrie. 1960. Principles and procedures of statistics. McGraw-Hill, New York. 481 p.

10. USDA Forest Service. 1969. California tree seed zone map, scale 1:1,000,000. Pacific Southwest Region, San Francisco. 1973. Oregon tree seed zone map, scale 1:500,000. Pacific Northwest Region, Portland.

11. Zaerr, J. B., B. D. Cleary, and J. L. Jenkinson. 1981. Scheduling irrigation to induce seedling dormancy. _In_ USDA Forest Serv. Gen. Tech. Rep. INT-109. p. 74-79. Intermountain Forest and Range Exp. Stn., Forest Serv., U.S. Dep. Agric., Ogden, Utah.

7. PHYSIOLOGY RESEARCH MADE FORESTATION WITH CONTAINER-GROWN SEEDLINGS SUCCESSFUL

R. W. TINUS and P. W. OWSTON

Plant Physiologist, Rocky Mountain Forest and Range Experiment Station, Forestry Sciences Laboratory, Flagstaff, Arizona 86001 and Plant Physiologist, Pacific Northwest Forest and Range Experiment Station, 3200 Jefferson Way, Corvallis, Oregon 97331

ABSTRACT

   Seedling physiology research in the past 15 years has improved production and performance of container seedlings. Biologically suitable containers have been developed, stages of growth are better understood, and suitable growing conditions for each stage have been determined. Container seedlings have increased survival and initial growth in many plantings and widened planting windows. Rapid production permits shorter planning horizons, resulting in less waste of seedlings and site preparation. Because container technology is readily transferable, container nurseries can be located almost anywhere. Ongoing research with great promise includes precise spacing of seedlings through sowing germinated seed, stress testing to determine planting stock quality, relating different growth aspects to each other to develop the best "balanced" seedling, and seeking growth control mechanisms at the cellular and biochemical levels.

7.1 INTRODUCTION

   For hundreds of years, foresters have been raising tree seedlings in outdoor, bare-root nurseries and outplanting the seedlings in the field with reasonable success. In the 1950's and 1960's, however, interest in raising and outplanting container-grown seedlings on a large scale arose in North America and elsewhere to shorten the production period and make more efficient use of genetically improved seed. Initially,

Duryea, M.L. and Brown, G.N. (eds.). Seedling physiology and reforestation succes
©1984, Martinus Nijhoff/Dr W. Junk Publishers, Dordrecht/Boston/London. ISBN 978-94-009-6139-5

the emphasis was on the automation and acceleration of production and planting, and there were some spectacular biological failures. More information was needed about seedling physiology and control of seedling growth in order for container seedlings to perform well. Furthermore, physiologists recognized the opportunity to utilize controlled environments for growing seedlings adapted to specific sites.

In the 1970's, intensive research on seedling physiology made possible reliable container seedling production and routine field success. Much of the research is reported or described in several manuals and symposia proceedings on container technology (35, 36, 27, 13). Physiologists have also played a significant role as consultants for developing nurseries and planting programs.

7.2 EARLY PROBLEMS

The Ontario tube, a cylindrical overlapping wrap of hard plastic filled with peat, was one of the first production-oriented containers. Seedlings were grown in a closed bottom tray and were outplanted with the container. Production of seedlings reached 22 million in 1967, and then dropped to less than 4 million in 1971 because of heavy losses in plantation: (14). Later study found that first, the containers were too small for growing vigorous seedlings; second, a container with a smooth vertical wall permits roots to circle; and third, a nondegradable container must be removed from the root ball before planting to allow root growth into surrounding soil.

Another example of a poorly-conceived container is the perforated polyethylene bag. This has been a favorite in developing countries because of low cost. However, it permits root circling. When seedlings are grown to considerable size, the circling roots form a weak spot at the base of the tree, and the tree may blow down or crack off. Plantation losses from this cause have been large in _Pinus_ and _Eucalyptus_ (3, 33), although other genera have had little problem from this (8, 25).

Seedlings produced in the early years of the technology
were often small, weak, and poorly hardened. Little attention
was paid to differences in water relations, mineral nutrition,
and pH control between the soilless potting media used for
container seedlings and the native soils of bare-root nurseries.
Later, lush seedlings were being produced in modern greenhouses
with little or no concern for acclimation to the light,
temperature, and moisture conditions on the planting site.

7.3 DEVELOPMENT OF SUITABLE CONTAINERS

A major improvement was the development of biologically
suitable containers. The small size of early containers
resulted in seedlings with relatively large tops and small
roots. Furthermore, it was soon realized that high bed
densities resulted in spindly seedlings, and that large
caliper was essential to reliable field success (31). Most
containers now used operationally range from 65 to 165 cc in
volume and are grown at a density of 6 to 11 per dm^2.

The shape of the container was equally important, because
it was found that vertical ribs, lack of sharp horizontal
corners, and a hole in the bottom for root egress and air
pruning would prevent root spiraling (15). A newer method
uses copper carbonate in latex paint to chemically prune the
roots as they contact the container wall (6, 19).

As soon as the seedling is outplanted, its roots must
quickly grow into the surrounding soil. Therefore, any
container that obstructs root growth must be removed from the
root ball before planting. This lesson took a long time to
learn, because it is so convenient operationally to plant the
container with the seedling (1, 32). Concern for root growth
of container seedlings led to an international symposium on
root form of all types of planted trees (38). All of the
many types of containers now on the market that are not
biodegradable or designed to be penetrated by roots have the
same features that make them biologically acceptable, and the
container is always removed before outplanting.

7.4 GROWING CYCLES

In an outdoor nursery, the weather determines the growing cycle. In greenhouses, however, where most container seedlings are grown, the cycle is not only controllable, but seedlings must be made to progress through it. The advantages are that accelerated growth and growth at other-than-normal times of the year are possible.

During its growth in the nursery, a seedling passes through four, five, or six distinct stages of growth, depending on where and when it is to be outplanted (35). Each stage has specific optimum conditions which vary with species and ecotype. Although physiologists have known all this in general, it has taken more than a decade to elucidate seedling requirements in sufficient detail to permit reliable production and flexible growing schedules.

Rapid, uniform germination is critical in a container nursery because of the relatively short growth cycle and rapid growth rates that tend to leave late germinants in a suppressed condition. Some seed treatments used in bare-root nurseries to insure rapid and complete germination (37), such as fall sowing to permit natural stratification, are not suitable for the container nursery. Others, while adequate for a bare-root nursery, are marginally adequate in a container nursery, where differences in germination time of a few days are important. Also, the ability to establish exactly one seedling per cavity without excessive thinning or transplanting is critical. This has forced a reexamination of seed preparation and germination procedures for container sowings that were thought to be adequate for bare-root nurseries (34).

During the period of rapid initial growth, the seedling grows exponentially in height and dry weight, as long as conditions remain near optimum. It is important that such growth be maintained until the seedling is as tall as desired for two reasons. First, in many species and ecotypes, height growth is hard to restart. Second, because of the exponential nature of the growth, a small difference in growing time results in a disproportionately large difference in final tree size. Either way, valuable time is lost, and the crop may not meet desired specifications.

Under greenhouse conditions, growth patterns and morphology may be quite different than they are in the outdoor nursery. This has prompted a number of studies to determine what constitutes normal growth under forced conditions (24, 30).

Of all the growth stages, dormancy induction and cold hardening have been least understood by growers (7). Whereas germination and growth are easily observed and measured, visible changes during dormancy induction are subtle and may be totally absent during hardening. Because container seedlings are almost always grown in a greenhouse, or at least under a cover that affects light, temperature, and air movement, seedlings usually must be acclimated to the outside and programmed physiologically to perform well when outplanted. The opportunity to do this is far greater in the greenhouse container nursery than in the outdoor bare-root nursery. The necessity to take deliberate action to harden seedlings is also greater, because nature will not do it. Too many poorly hardened trees are still being produced. Part of the problem is lack of appreciation for the need, and part is inability to identify and measure important physiological characteristics quickly enough to make day-to-day management decisions. Progress in learning the details of the hardening process, how to measure it, and how to optimize it has lagged behind progress in other areas.

Once full dormancy is attained, nursery managers holding seedlings for spring planting must understand specific chilling requirements for vigorous spring growth. Nurseries in mild climates growing seed sources from cold-climate sites may have to use cold storage facilities to fulfill chilling and maintain dormancy until planting.

7.5 CULTURAL PRACTICES

One of the principal features distinguishing the greenhouse container nursery from the outdoor bare-root nursery is the increased control of environmental conditions. Optimum

temperatures for rapid growth are now well known for many
species (35), and season-long effects of different types of
temperature regimes have been studied for a few species (22).
Optimum temperatures for budset and hardening are less well
known.

Irrigation practices are important for conditioning as
well as for growth. Seedlings grown under heavy irrigation
may well be large, but they are more susceptible to frosts
and drought after outplanting because of lush foliage and
high ratios of tops to roots (23). A variety of methods have
been developed to measure and maintain proper moisture
relations. Measurement techniques include weighing containers,
the pressure chamber, bimetal probe, tensiometer, and educated
finger. Each has its own advantages with respect to speed,
accuracy, and suitability on different sizes of containers
and seedlings (18).

Proper moisture relations critically depend on use of a
growing medium with the correct balance and amount of water
retention capacity and airspace. For these and other reasons,
peat alone or mixed with vermiculite, perlite, and sometimes
other components has been the medium of choice. However, an
increasing number of container nurseries are far from suitable
sources of peat. Properly composted, a wide variety of
locally available organic materials should be acceptable
substitutes (4, 28, 29). In the past, lack of proper composting
has resulted in many poor crops or complete failures. Inadequate
attention to the pH of the potting mixture has also caused
serious problems, but prevention and curative procedures are
known.

Optimum mineral nutrient concentrations have been studied
in hydroponic culture or pot tests by many workers during the
past 30 years. These tests have been of limited usefulness
for predicting response to fertilizer of large trees in the
field or seedlings in the outdoor nursery because of complex
soil chemistry and limited control of the water supply. With
the development of the greenhouse container nursery, however,
knowledge of optimum nutrient concentrations has become very

valuable. This results from using a completely soluble fertilizer and a growing medium that has a minimal effect on nutrient ion concentrations, and the ability to change nutrients by leaching.

One of the unique advantages of the greenhouse nursery is its capability of producing crops any time of the year. For most woody species, this would be impossible without control of photoperiod. The phenomenon of control of dormancy by day length was discovered in the 1920's (11), and the mechanism was described in the 1950's (5). By the 1960's, lights at night came into common use in horticultural greenhouses to control flowering. Light at night was tested in bare-root nurseries to increase growth, but without also controlling the temperature, it is not surprising that it failed (2, 26). In contrast, extended photoperiod in the greenhouse nursery has become the principal means by which shoot growth is maintained and dormancy is prevented. Intermittent lighting systems have been developed which are more energy efficient and permit more stem elongation than continuous light. There is now sufficient information on intensity, duration, and quality of light required for many species (35).

Shortening the photoperiod to promote bud dormancy and cold hardening has been used successfully. However, because the mechanics of deploying and retracting blackout cloth daily are cumbersome, the technique has not been widely used.

Seedlings grown in bare-root nurseries are usually mycorrhizal. Only when nurseries have been newly established in dry climates, or too heavily fumigated, has there been a problem. However, when seedlings began to be grown on a relatively sterile medium that was changed each crop, mycorrhizal fungi usually did not become established. Research was focused on this problem, and successful inoculation techniques were developed (20, 17).

7.6 IMPACTS ON FORESTATION

Container seedlings often survive better and grow more the first season than do bare-root seedlings of the same size.

Smaller container seedlings may survive as well as much
larger bare-root seedlings and may be easier and faster to
plant--especially on rocky or debris-covered sites.

Recent economic analyses indicate that for the same or
better survivability and growth, container seedlings may be
as inexpensive as bare-root seedlings (12). One of the
biggest advantages is that rapid and off-season production
permits shorter planning lead times, resulting in less waste
of seedlings and site preparation.

As the forestation job becomes larger, it is increasingly
more difficult to complete programs within the traditional
bare-root planting season. Container seedlings offer an
opportunity to plant whenever the ground is not frozen and
soil moisture is adequate. In parts of the South, that may
be all year. Almost everywhere, it represents a wider planting
window than is available with bare-root stock.

Certain species are more easily grown in containers than
as bare-root seedlings. Western hemlock (Tsuga heterophylla
(Raf.)) and firs (Abies spp.) are good examples (21, 22).
Container-grown grass plugs have been found to be very effective
in stabilizing disturbed land where sod is too expensive and
seed would be washed away. Establishment of shrubs and forbs
for wildlife habitat is another use of container technology
(10).

Bare-root nurseries have exacting soil requirements that
make it difficult to find suitable nursery sites. Many
nurseries are in less-than-ideal locations that require site-
specific research to make possible reliable production of
high quality seedlings. In contrast, greenhouse container
nurseries can be located almost anywhere, and the technology
is easily transferable.

Container technology is suited to development of complete
forestation systems, as exemplified by the dibble-tube system
developed for Hawaii (39).

Finally, the research and technology pursued to make the
greenhouse container nursery a success are also being applied

to improve the reliability of high quality seedling production
in the bare-root nursery.

7.7 PROSPECTS FOR THE FUTURE

The container nursery has demonstrated the value of precise
square or hexagonal spacing of seedlings, which results in
very uniform growth and a low cull factor not matched by row-
seeded beds. The need to sow an exact number of seed in a
well-defined geometric pattern has resulted in the development
of truly precise seeders, such as the vacuum plate and
shutter box.   Their use is now spreading to the bare-root
nursery. At present, the ability to establish exactly one
seedling per cavity is limited only by the quality of the
seed sown. Research on practical methods for sowing germinated
seed is underway and promises to remove the last barrier to
perfect spacing without thinning or transplanting, both of
which are expensive, and thinning wastes seed.

As a result of increased ability to control growing conditions
and increased interest in making certain that forestation is
successful, research is underway to develop methods to measure
the physiological condition of both bare-root and container
seedlings to predict their responses in the field. Drought
stress test results at Oregon State University have been
found to correlate well with field performance (9). Testing
for root regenerating potential has become the accepted means
to determine suitable lifting dates at bare-root nurseries
(16). This has to be done by seed origin and at almost every
nursery site--another cost for developing bare-root nurseries
that is not necessary for container nurseries.

Several tests for cold hardiness are available. Whole
plant freezing is accurate and foolproof, but takes 1 week or
more. Electrolyte leakage is also workable and can be done
in 48 hours. So far, differential thermal analysis is the
quickest (30 minutes) and easiest, but is not usable on many
species, including pines. The search for rapid, simple, and
accurate tests continues, because these are needed not only

to facilitate research, but to permit day-to-day management decisions.

Because physiological research has improved forestation success, there is now more interest in pursuing basic research whose payoff is likely to be long-term. One direction is learning how the different aspects of growth are related. Perhaps one day the characteristics of a "well-balanced, high quality" seedling will be measurable quickly and objectively. Another direction is to seek growth control mechanisms at the cellular and biochemical levels. Much progress has already been made. For example, tissue culture is already in large-scale use to propagate many species. When tissue culture is used, the greenhouse container nursery is certain to be intermediate between the flask and the field.

REFERENCES

1. Arnott, J. T. 1973. Evolution of the styroblock reforestation concept in British Columbia. Commonwealth For. Rev. 52(1):72-78.
2. Bean, D. 1964. Artificial lighting fails to stimulate height growth of white pine seedlings in nursery studies. Tree Planters Notes 64:23-26.
3. Ben Salem, B. 1971. Root strangulation--A neglected factor in container grown nursery stock. M.S. thesis, 50 p. Univ. of Calif., Berkeley.
4. Blumenthal, Susan G., and Donald E. Boyer. 1982. Organic amendments in forest nursery management in the Pacific Northwest. USDA For. Serv. Adm. Rep., 71 p. Pac. Northwest For. and Range Exp. Stn., Portland, Oreg.
5. Borthwick, H. A., and H. M. Cathey. 1962. Role of phytochrome in control of flowering in chrysanthemums. Bot. Gaz. 123(3):155-162.
6. Burdett, A. N., and P.A.F. Martin. 1982. Chemical root pruning of coniferous seedlings. HortScience 17(4):622-624.
7. Carlson, W. C., W. D. Binder, C. O. Feenan, and C. L. Preisig. 1980. Changes in mitotic index during onset of dormancy in Douglas-fir seedlings. Can. J. For. Res. 10:371-378.
8. Carlson, W. C., C. L. Preisig, and L. C. Promnitz. 1980. Comparative root system morphologies of seeded-in-place, bareroot, and container-cultured plug Sitka spruce seedlings after outplanting. Can. J. For. Res. 10:250-256.
9. Duryea, M. 1983, in press. The nursery technology cooperative: A coordinated effort to improve seedling quality. Proc. Intermt. nurserymen's meet. Las Vegas, Nev., Aug. 9-11, 1983.

10. Ferguson, Robert B., and Stephen B. Monsen. 1974. Research with containerized shrubs and forbs in southern Idaho. p. 349-358. In Proc. No. Amer. containerized forest tree symp. R. W. Tinus, W. I. Stein, and W. E. Balmer, eds. Great Plains Agric. Counc. Publ. 68.

11. Garner, W. W., and H. A. Allard. 1923. Further studies in photoperiodism: The response of the plant to relative length of day and night. J. Agric. Res. 23:871-920.

12. Guldin, R. W. 1982. Capital intensity and economics-of-scale for different types of nurseries. p. 87-96. In Proc. South. containerized forest tree seedling conf. R. W. Guldin and J. P. Barnett, eds. Savannah, Ga., Aug. 25-27, 1981. USDA For. Serv. Gen. Tech. Rep. SO-37, 156 p. South. For. Exp. Stn., New Orleans, La.

13. Guldin, R. W., and J. P. Barnett, eds. 1982. Proc. South. containerized forest tree seedling conf. Savannah, Ga., Aug. 25-27, 1981. USDA For. Serv. Gen. Tech. Rep. SO-37, 156 p. South. For. Exp. Stn., New Orleans, La.

14. Heeney, C. J. 1982. The status of container planting programs in Canada, 5. Ontario. p. 35-40. In Proc. Can. containerized tree seedling symp. J. B. Scarratt, C. Glerum, and C. A. Plexman, eds. Toronto, Ont., Sept. 14-16, 1981. Canada-Ontario Joint Forestry Research Committee Symp. Proc. O-P-10.

15. Hiatt, H. A., and R. W. Tinus. 1974. Container shape controls root system configuration of ponderosa pine. p. 194-196. In Proc. No. Amer. containerized forest tree seedling symp. R. W. Tinus, W. I. Stein, and W. E. Balmer, eds. Great Plains Agric. Counc. Publ. 68.

16. Jenkinson, James L. 1984, in press. Utilizing seed source lifting windows to improve plantation establishment of Pacific slope Douglas-fir. In New forests for a changing world, Proc. 1983 Society of American Foresters National Convention, Portland, Oreg., Oct. 16-20, 1983.

17. Marx, D. H., and J. P. Barnett. 1974. Mycorrhizae and containerized forest tree seedlings. p. 85-91. In Proc. No. Amer. containerized forest tree seedling symp. R. W. Tinus, W. I. Stein, and W. E. Balmer, eds. Great Plains Agric. Counc. Publ. 68.

18. McDonald, S. E., and S. W. Running. 1979. Monitoring irrigation in western forest tree nurseries. USDA For. Serv. Gen. Tech. Rep. RM-61, 8 p. Rocky Mtn. For. and Range Exp. Stn., Fort Collins, Colo.

19. McDonald, S. E., R. W. Tinus, and C.P.P. Reid. 1982. Root development control measures in containers: Recent findings. p. 207-214. In Proc. Can. containerized tree seedling symp. J. B. Scarratt, C. Glerum, and C. A. Plexman, eds. Toronto, Ont., Sept. 14-16, 1981. Canada-Ontario Joint Forestry Research Committee Symp. Proc. O-P-10.

20. Molina, R., and J. M. Trappe. 1982. Ch. 12. Applied aspects of ectomycorrhizae. In Advances in Agricultural Microbiology, N.S. Subba Rao, ed. Oxford and IBH Publishing Co., New Delhi. 305 p.

154

21. Owston, P. W., and T. T. Kozlowski. 1976. Effects of temperature and photoperiod on growth of western hemlock. In Proc. Western hemlock manage. conf. W. A. Atkinson and R. J. Zasoski, eds. College of Forest Resources, Univ. of Wash., Seattle, May 1976. 108 p.

22. Owston, P. W., and T. T. Kozlowski. 1981. Growth and cold hardiness of container-grown Douglas-fir, noble fir, and Sitka spruce seedlings in simulated greenhouse regimes. Can. J. For. Res. 11:465-474.

23. Owston, P. W., and W. I. Stein. 1977. Production and use of container seedlings in the West. In Proc. Intermt. forest nurserymen's assoc. W. L. Louchs, ed. Kansas State University, Manhattan. 117 p.

24. Powell, G. R. 1983. A comparison of early shoot development of seedlings of some trees commonly raised in the northeast of North America. p. 1-24. In Proc. Northeast. area nurserymen's conf. Halifax, Nova Scotia, July 25-29, 1982. Nova Scotia Dept. Lands and Forests, Truro, N. S.

25. Preisig, C. L., W. C. Carlson, and L. C. Promnitz. 1979. Comparative root system morphologies of seeded-in-place, bareroot, and containerized Douglas-fir seedlings after outplanting. Can. J. For. Res. 9:399-405.

26. Read, R. A., and W. T. Bagley. 1967. Response of tree seedlings to extended photoperiods. USDA For. Serv. Res. Pap. RM-30, 16 p. Rocky Mtn. For. and Range Exp. Stn., Fort Collins, Colo.

27. Scarratt, J. B., C. Glerum, and C. A. Plexman, eds. 1982. Proc. Can. containerized tree seedling symp. Toronto, Ont., Sept. 14-16, 1981. Canada-Ontario Joint Forestry Research Committee Symp. Proc. O-P-10, 460 p.

28. Solbraa, Knut. 1979. Composting of bark. III. Experiments on a semi-practical scale. Medd. Norsk Inst. Skogforsk 34:387-439.

29. Solbraa, Knut. 1979. Composting of bark. IV. Potential growth-reducing compounds and elements in bark. Medd. Norsk Inst. Skogforsk 34:443-508.

30. Thompson, S. 1981. Shoot morphology and shoot growth potential in 1-year-old Scots pine seedlings. Can. J. For. Res. 11:789-795.

31. Tinus, R. W. 1974. Characteristics of seedlings with high survival potential. p. 276-282. In Proc. No. Amer. containerized forest tree seedling symp. R. W. Tinus, W. I. Stein, and W. E. Balmer, eds. Great Plains Agric. Counc. Publ. 68.

32. Tinus, R. W. 1974. Large trees for the Rockies and Plains. p. 112-118. In Proc. No. Amer. containerized forest tree seedling symp. R. W. Tinus, W. I. Stein, and W. E. Balmer, eds. Great Plains Agric. Counc. Publ. 68.

33. Tinus, R. W. 1978. Root form: What difference does it make? p. 11-15. In Proc. Root form of planted trees symp. E. Van Eerden and J. M. Kinghorn, eds. Victoria, B.C., May 16-19, 1978. British Columbia Ministry of Forests/Can. For. Serv. Jt. Rep. 8.

34. Tinus, R. W. 1982. Effects of dewinging, soaking, stratification, and growth regulators on germination of green ash seed. Can. J. For. Res. 12(4):931-935.

35. Tinus, R. W., and S. E. McDonald. 1979. How to grow tree seedlings in containers in greenhouses. USDA For. Serv. Gen. Tech. Rep. RM-60, 256 p. Rocky Mtn. For. and Range Exp. Stn., Fort Collins, Colo.

36. Tinus, R. W., W. I. Stein, and W. E. Balmer, eds. 1974. Proc. No. Amer. containerized forest tree seedling symp. Great Plains Agric. Counc. Publ. 68, 458 p.

37. U.S. Department of Agriculture, Forest Service. 1974. Seeds of woody plants in the United States. Agriculture Handbook 450, 883 p. Washington, D.C.

38. Van Eerden, E., and J. M. Kinghorn, eds. 1978. Proc. Root form of planted trees symp. Victoria, B.C., May 16-19, 1978. British Columbia Ministry of Forests/Can. For. Serv. Jt. Rep. 8. 357 p.

39. Walters, G. A. 1981. Why Hawaii is changing to the dibble-tube system of forestation. J. For. 79(11):743-745.

8. RELATING SEEDLING PHYSIOLOGY TO SURVIVAL AND GROWTH IN CONTAINER-GROWN SOUTHERN PINES

J. P. BARNETT

Principal Silviculturist, Southern Forest Experiment Station, USDA Forest Service, 2500 Shreveport Highway, Pineville, LA 71360

ABSTRACT

A variety of physiological and morphological charac-teristics of container grown southern pine seedlings were measured at the time of outplanting and then related to field performance. The variables measured included: needle chlorophyll content, degree of mycorrhizal inoculation, biochemical constituents, seedling height and stem diameter, dry weight of roots and shoots, and shoot/root ratios. Although many of the variables were significantly correlated to one another, seedling heights were most positively related to growth in the field. Most of the characteristics measured could not be closely related to field survival, because sur-vival was unusually high. Cultural treatments applied during the greenhouse growing period affected both initial seedling development and field performance.

8.1 INTRODUCTION

Using physiological qualities of seedlings to estimate field performance has long been recognized as desirable. Wakeley (19), in discussing physiological quality of nursery stock at a meeting 35 years ago, considered physiological quality to be a much sounder guide to seedling performance than deficiencies in morphological grade. Wakeley also cited evidence that morphological grades and physiological qualities do not necessarily coincide, nor are they necessarily iden-tical with the capacity to survive and grow. Although Wakeley's comments were directed at conditions that occurred in the production of bare-root nursery seedlings, they apply

Duryea, M.L. and Brown, G.N. (eds.). Seedling physiology and reforestation succes
©1984, Martinus Nijhoff/Dr W. Junk Publishers, Dordrecht/Boston/London. ISBN 978-94-009-6139-5

as well to containerized seedling production.

Containerized seedlings can survive in the field better than nursery-grown seedlings, even when they are of poor quality or low vigor, as long as soil moisture and other environmental conditions at the time of outplanting are near optimal. This reflects the planting of intact versus partial root systems of the two types of seedlings (2). However, when less favorable conditions are met soon after outplanting, the morphological and physiological conditions of the seedlings determine their ability to survive and grow. Several workers have noted that container-grown seedlings that are large and woody survive and grow better than smaller seedlings on difficult sites or where competition is severe (1, 7, 9, 22). Southern pine *(Pinus* spp.) seedlings are usually large enough at 12 to 14 weeks to perform well in the field. Age alone, however, is not a reliable criterion of when to plant, because seedling development varies greatly by season, facility, and cultural treatment.

Wakeley's (20) grading system for bare-root seedlings, based primarily on seedling height, diameter, and nature of the stem, is still the best available for southern pines. Because of differences in age, development, and cultural regimes between bare-root and container-grown seedlings, grades established for bare-root may not be the same as for those grown in containers. Over the past several years, physiological qualities of container-grown seedlings that relate to seedling morphology and field performance have been evaluated. Data from those tests will be discussed in this paper.

8.2 MORPHOLOGICAL CHARACTERISTICS

8.2.1 Development of varying seedling morphologies

Selected morphological and physiological characteristics developed by varying seedling age, light, and nutrient conditions were related to seedling performance in the field. Four 25-seedling plots each of loblolly (*P. taeda* L.), longleaf *(P. palustris* Mill.), and shortleaf (*P. echinata* Mill.) pine, were used to evaluate field response to initial

seedling condition. Seedlings were grown in either
Styroblock-2® [1] (40 cc) or Styroblock-4® (65 cc) containers
filled with a 1:1 peat-vermiculite medium for 10, 12, and 14
weeks.

Commercial water soluble fertilizers were applied weekly
through the water system, beginning 1 week after seed germination. Peters® 20-19-18 NPK was typically used at a
concentration of 150 ppm of nitrogen. Supplemental light
was provided to some treatments by a bank of eight
fluorescent lamps about 45 cm above the seedlings. A
16-hour photoperiod was maintained for these treatments.

Loblolly and shortleaf seeds were given 30 days of moist
prechilling prior to sowing. Longleaf seeds were sown
without treatment.

Prior to outplanting in the field, four 5-seedling replications for each tree species were measured to determine top
height (loblolly and shortleaf only), stem diameter, and
stem and root dry weight.

8.2.2 Seedling characteristics

A variety of morphological characteristics have been
directly related to either survival or growth of seedlings
in the field. However, in a number of outplantings survival
was excellent regardless of seedling characteristics. This
made it impossible to develop consistent relationships among
seedling grades and field survival. The range in values of
morphological characteristics shown in Table 1 is not
necessarily optimum for consistent field performance. The
level of seedling quality that will provide acceptable performance when there are good environmental conditions may
not be adequate under highly stressful conditions.

Containerized seedlings should survive better than
bareroot nursery stock on difficult sites, but early growth
after outplanting may be more indicative of seedling quality

---

[1] Mention of trade names is for information only and
does not constitute endorsement by the USDA Forest Service.

than survival. In our tests, large seedlings consistently grew better after outplanting than small ones. Seedlings whose morphological values are in the higher ranges (Table 1), are more likely to succeed in the field.

Table 1. Ranges in morphological values associated with acceptable field success of container-grown southern pine seedlings

Morphological characteristics	Species		
	Loblolly pine	Longleaf pine	Shortleaf pine
Height (mm)	90 - 318	-------	59 - 161
Shoot diameter (mm)	1.5 - 3.0	1.9 - 3.6	1.2 - 1.8
Shoot weight (mg)	228 - 1244	336 - 1352	113 - 440
Root weight (mg)	90 - 189	61 - 190	35 - 178
Shoot/root ratios	2.0 - 6.7	1.9 - 8.7	1.4 - 6.9

8.3 PHYSIOLOGICAL CHARACTERISTICS
8.3.1 Physiological characteristics evaluated

Seedlings showing the varying morphological conditions described in table 1 were also analyzed for several physiological characteristics. Analyses were made of chlorophyll content of the needles (23) and of total and reducing sugars (14, 15), nitrogen contents (10), and lipids (6) of the stems and roots. Determinations of the amount of mycorrhizal inoculation on root systems were made following techniques reported by Barnett (4) and Ruehle et al. (16).

Chlorophyll contents of needles have been related to seedling quality by other studies (18). Sugars, lipids, and nitrogen (protein) are active metabolic constituents in plants and in the past have been related to various aspects of seedling development.

8.3.2 Variability of physiological characteristics

The ranges in values obtained from the analyses of physiological properties are listed in Table 2. Since overall

field survival was excellent, no correlations between these
values and survival were obtained, but growth related to certain of these physiological properties.

Table 2. Ranges in physiological values associated with
acceptable field sucess of container-grown southern pine
seedlings

Physiological characteristics	Species		
	Loblolly pine	Longleaf pine	Shortleaf pine
---------------Shoot---------------			
Lipids (%)	8.6 - 12.6	8.3 - 13.8	-----
Total sugars (%)	1.5 - 2.2	2.5 - 4.6	-----
Reducing sugars (%)	.9 - 1.6	1.4 - 3.4	-----
Nitrogen (%)	2.3 - 3.3	2.8 - 3.4	-----
Chlorophyll (mg/g)	0.9 - 1.7	0.9 - 2.0	0.7 - 1.4
---------------Root---------------			
Lipids (%)	4.2 - 4.9	4.2 - 6.3	-----
Total sugars (%)	1.9 - 2.7	2.6 - 4.3	-----
Reducing sugars (%)	.8 - 1.4	.9 - 2.3	-----
Nitrogen (%)	1.9 - 2.4	2.3 - 2.8	-----
Mycorrhizae (% Pt)	-------	0 - 16	0 - 23

8.4 RELATING SEEDLING PROPERTIES TO FIELD PERFORMANCE

The results of several analyses were combined in order to
obtain a more complete evaluation of how seedling physiology
and morphology relate to field performance. Since determinations of physiological characteristics were not made at
each analysis, comparisons of initial morphological and physiological characteristics were made to develop relationships between the two types of seedling performance.

8.4.1 Relating morphology and physiology

8.4.1.1 Biochemical constituents. Analyses of lipids,
total and reducing sugars, and nitrogen were run for
longleaf and loblolly pine seedlings. Shortleaf seedlings
were not available for these analyses. Correlations of the

data from those analyses were made with seedling morphological characteristics and with survival and growth after outplanting. In this test, field survival was excellent, greater than 95 percent, regardless of treatments. Also, after 1 year in the field, no differences in growth measurements due to treatment occurred. Because of the excellent field performance, obviously the biochemical constituents could not be related to survival or growth.

The data did allow comparisons among morphological and physiological characteristics. Results from these correlations provided some interesting results (Table 3). Responses differed greatly by species. No statistically significant correlation coefficients occurred between biochemical constituents and morphological properties in loblolly pine. However, in longleaf pine significant relationships were found. Lipids and both total and reducing sugars were correlated to seedling size. Positive coefficients of 0.8 or more occurred between stem diameters and weights and sugars. Larger seedlings had greater amounts of sugars per unit weight than smaller seedlings. Conversely, lipids were negatively related to seedling size. Large seedlings had less lipids and more sugars available for growth processes. Longleaf root weights were positively related to both total and reducing sugars. But no relationships occurred between lipids and root weights.

8.4.1.2 Chlorophyll. Although needle chlorophyll contents were not related to initial morphological characteristics in longleaf pine, they are positively correlated to stem size in loblolly pine (Table 3). In this test different nutrient regimes were used during the greenhouse culture period, and thus differences in seedling quality may have been due to a close relationship between chlorophyll and nitrogen contents. Unfortunately, these analyses were not made on the same individual seedlings, so direct comparisons cannot be made.

Table 3. Summary of correlation coefficients relating initial morphological and physiological characteristics.

Species	Morphological characteristic	Lipids	Total sugar	Reducing sugar	Nitrogen	Chlorophyll
		-----Correlation coefficients-------				
Long-leaf	Stem diameter	-.624*[1]	.818*	.838*	-.420	-.040
	Shoot dry wt.	-.655*	.857*	.857*	-.429	.377
	Root dry wt.	-.195	.736*	.694*	-.292	-.412
Loblolly	Height	.177	-.003	-.163	-.496	.687*
	Stem diameter	.146	.149	-.054	-.423	.764*
	Shoot dry wt.	.247	.161	-.108	-.394	.572*
	Root dry wt.	-.079	-.091	.036	-.568	-.313

[1]/An asterisk represents statistical significance at the 0.05 level.

8.4.1.3 Mycorrhizae. Although the presence of ectomycorrhizae on seedling root systems is not a physiological condition, it is a morphological adaptation of the root system that allows the roots to function of a different physiological manner. Since the presence of mycorrhizae is not usually considered a morphological trait, I have compared its presence with the growth of longleaf and shortleaf pine seedlings. Results from these evaluations showed no relationship between seedling development in the greenhouse and amount of mycorrhizae on the root systems (4).

8.4.2 Relating field performance to seedling traits.

One problem in relating seedling cultural treatments in the greenhouse to conditions encountered after outplanting is the lack of control of environment conditions once seedlings are planted in the field. It may be necessary to rerun the same study for several years to obtain realistic evaluations of the treatment responses. Such a problem occurred in the conduct of our studies. The tests in which many of the physiological characteristics were evaluated showed excellent survival and growth in the field--there were no differences due to these varying seedling conditions. These tests were rerun several times, but physiological analyses were not made in each repeat test. It became necessary, then, to correlate the physiological traits measured to common morphological characteristics determined in other tests that could be related with field performance.

The morphological characteristics that are directly related to field performance in at least one test include: stem height, stem diameter, dry weight, shoot/root ratios, mycorrhizal development, and presence of secondary needles. Chlorophyll content was a physiological trait measured in additional tests. The value of any of these measurements as indicators of performance for container-grown seedlings varies greatly.

8.4.2.1 <u>Secondary needles.</u> The development of fascicle or secondary needles is one criterion used by Wakeley (20) in his seedling grading system for nursery stock. Our tests with container-grown southern pines agree that the presence of secondary needles is an important indicator of seedling development (3). Secondary needles develop when the stem becomes woody and stiff. This condition represents a stage when the seedlings become more hardy and less susceptible to cold and drought damage. Thus, seedlings that have secondary needles are more vigorous than those that have not yet reached this stage of development.

8.4.2.2 <u>Chlorophyll content.</u> Chlorophyll content in seedling needles has been shown to give an estimate of stock quality (18). In one of our tests, the needle chlorophyll content of seedlings at planting was correlated to their height 1, 2, and 3 years later (Table 4). In this particular study, high chlorophyll content was related to seedling vigor. Chlorophyll is generally a nonspecific indicator that is influenced by many factors. When seedlings are grown with an abundance of nutrients, chlorophyll content will probably not be closely related to field performance, because all will likely have high chlorophyll levels.

Table 4. Summary of correlation coefficients relating initial seedling development to field performance (test 1).

Seedling[1] characteristics	Survival Feb. 1977	Feb. 1978	Feb. 1979	Height Feb. 1977	Feb. 1978	Feb. 1979
----------Correlation coefficients---------						
Loblolly Pine [2]						
Height	0.021	0.238	0.324	.972*	0.743*	0.738*
Stem diameter	.168	.361	.455	.913*	.819*	.833*
Root weight	-.345	-.259	-.339	.078	-.225	-.259
Stem weight	.049	.294	.315	.870*	.646*	.655*
Chlorophyll	.235	.418	.595*	.771*	.915*	.899*
Longleaf Pine [3]						
Stem diameter	.079	.046	.020	.106	-.112	-.162
Root weight	.534*	.522*	.479	.396	-.435	-.431
Stem weight	-.252	.296	.176	.548*	.590*	.555*
Chlorophyll	-.418	-.566*	.443	.866*	.810*	.798*
Shortleaf Pine						
Height	.052	.147	.283	.929*	.829*	.784*
Stem diameter	-.142	-.058	.099	.761*	.760*	.687*
Root weight	.170	.157	.182	.061	-.021	-.057
Stem weight	.030	.128	.249	.773*	.743*	.666*
Chlorophyll	.281	.193	.302	.913*	.861*	.832*

[1] Seedlings outplanted June 22, 1976.
[2] An asterisk represents statistical significance at the 0.05 level.
[3] Longleaf growth was evaluated by measuring root-collar diameter rather than height.

8.4.2.3 Shoot/root ratios. Seedlings usually have been reared with the view that growth and survival will be best when the shoot/root ratio is between 1 and 3 (8, 20). Recent work by Walker and Johnson (22) with northern species of spruce (*Picea* spp.) and pine shows that much higher

166

shoot/root ratios may be better for container-grown seedlings.
Regression analyses of their data indicate that the weight
obtained 1 year after planting is proportional to initial
seedling weight and shoot/root ratio; larger seedlings with
shoot/root ratios of up to 7.4 had significantly greater
weight increases than smaller seedlings with ratios of 2.0.
A similar relationship was found with southern pines when
shoot/root ratios were related to growth in the field (FIGURE
1).

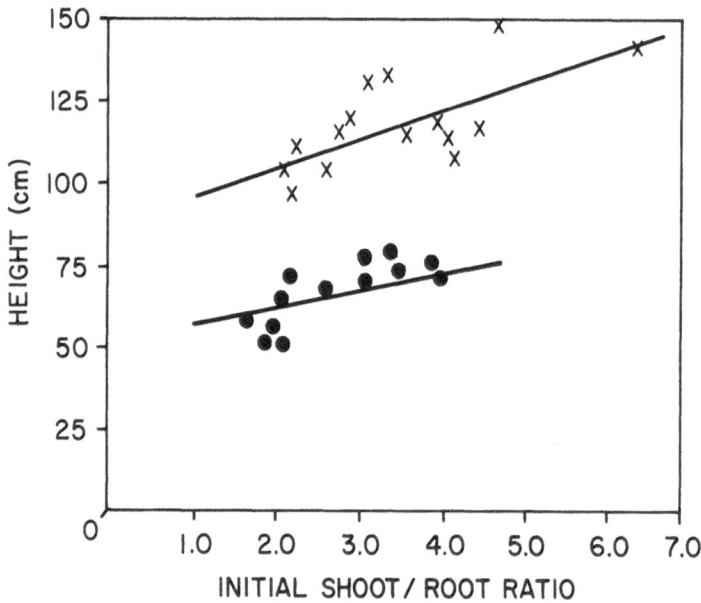

FIGURE 1. Initial shoot/root ratio and seedling height
relationships 2 1/2 and 3 1/4 years after outplanting, based
on two separate outplantings of loblolly pine.

It is apparent from the data shown in FIGURE 1 that a so-
called "balanced" seedling is not necessary or even desirable
with container-grown plants. In our work with container-
grown seedlings, higher shoot/root ratios are more a function
of larger shoots than variations in root size. Therefore, we
have concluded that shoot/root ratio is generally not a
meaningful criterion when evaluating containerized southern
pine seedlings.

8.4.2.4 <u>Mycorrhizae</u>. The visible presence of mycorrhizae
on bare-root slash and loblolly pine seedlings indicates that
there will be increased survival of the nursery stock (11,
17). The amount of mycorrhizae can have a significant effect
on survival and growth of southern pine nursery stock, but
container-grown plants may respond differently.

Inoculation of container-grown longleaf and shortleaf pine
seedlings with *Pisolithus tinctorius* (Pt) did not improve
seedling survival either in these tests or in an earlier study
where shortleaf seedlings inoculated with Pt and *Thelephora
terrestris* (Tt) were outplanted on a dry site in the Ouachita
Mountains in Arkansas (16). In this Pt test, growth was
affected by the treatments (FIGURE 2). Some Pt inoculation
treatments resulted in larger seedlings than the low fertility
control without Pt, but all were smaller than those in the
high fertility control without Pt. Basically, there was no
advantage in reducing fertility levels to insure that
increased amounts of mycorrhizae would develop (13). The
larger seedlings that were not inoculated, but received higher
fertility levels in the greenhouse, performed better on normal
reforestation sites.

Correlation coefficients between seedling characteristics
and field performance indicated that initial seedling size was
more an indicator of future growth than the amount of
mycorrhizae on the root system. This makes differences in
seedling size at the time of outplanting an important consideration for field performance.

8.4.2.5 <u>Stem diameter</u>. Stem diameter was a characteristic
closely related to seedling development in loblolly and
shortleaf pine (Table 4). Other tests confirmed the relationship of stem diameter of loblolly pine to seedling growth
after outplanting (Table 5). Loblolly stem diameters were
consistently correlated with survival. However, the combination of stem diameter with other easily measureable traits
should improve predictions of field performance.

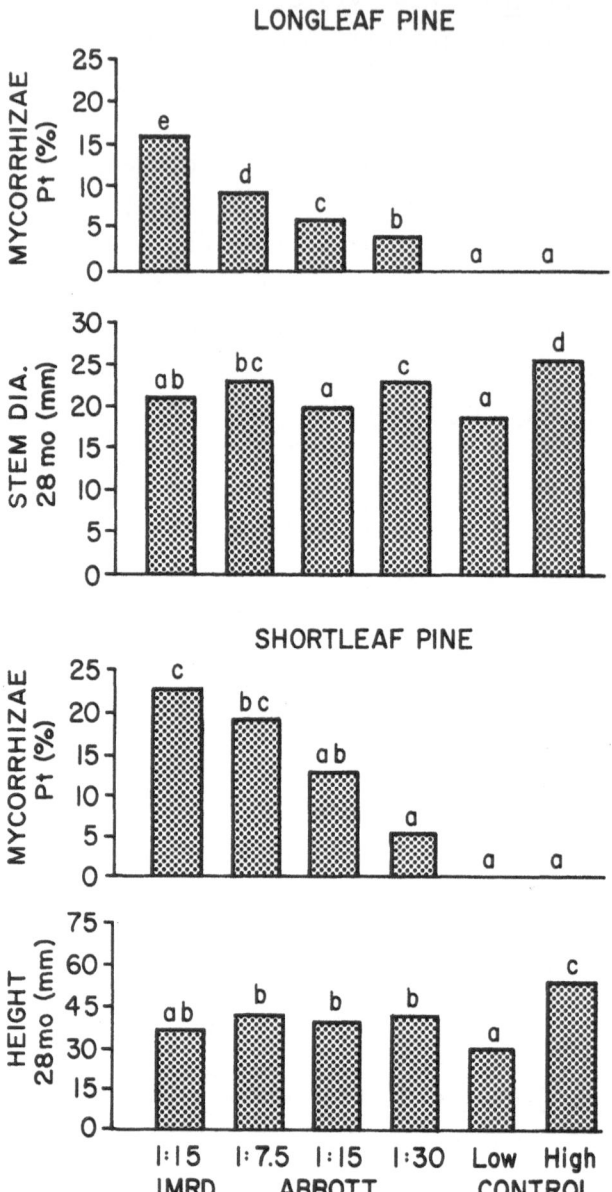

FIGURE 2. Longleaf and shortleaf seedling performance after outplanting as influenced by mycorrhizal inoculation treatments. Separately for each seedling characteristic, treatments with the same letter are not significantly different at the 0.05 level. Adapted from Barnett (4).

Table 5. Summary of correlation coefficients relating initial seedling development to field performance (test 2).

Seedling [1] characteristics	Survival			Height	
	March 1978	Feb. 1979	Jan. 1980	Feb. 1979	Jan. 1980
	───────Correlation coefficient──────────				
	Loblolly Pine [2]				
Height	0.441	0.497*	0.481	0.430	0.615*
Stem Diameter	.475	.532*	.522*	.496	.601*
Root weight	.519*	.527*	.514*	.368	.476
Shoot weight	.417	.470	.460	.432	.518*
	Longleaf Pine [3]				
Stem Diameter	.320	.317	.346	.670*	.510*
Root weight	.534*	.514*	.495	.196	.222
Shoot weight	.131	.149	.207	.292	.384

[1] Seedlings outplanted June 27, 1977.
[2] An asterisk represents statistical significance at the 0.05 level.
[3] Longleaf growth was evaluated by measuring root-collar diameter rather than height.

8.4.2.6 Dry weights. Dry weights of seedling stems at the time of outplanting were correlated with heights in the field over several years (Tables 4 and 5), but were not closely related to survival. There was no consistency in correlations of dry weight of roots at outplanting to survival in most instances. In test 1, root weight was related to field survival of longleaf pine only (Table 4). In test 2, correlations between root weight and survival occurred with both loblolly and longleaf pine (Table 5). In this latter study, initial root weights did not relate to seedling height increases. Differences in response among studies seem related to environmental conditions at or shortly after planting.

8.4.2.7 <u>Seedling heights.</u> The height of a seedling when
outplanted is generally a good indicator of subsequent
field performance (9, 22). Our studies show that this
holds true for container-grown southern pines (Tables 4 and
5). Not only is height at time of outplanting closely
related to subsequent heights, but it is also correlated to
incremental growth for a number of years (12). How long
this relationship will hold is open to question. Blair and
Cech's (5) work with slash pine (*Pinus elliottii* Engelm.)
nursery stock has shown that Wakeley's grade 1 and 2
seedlings produced significantly more volume after 13 years
than grade 3 seedlings (FIGURE 3). In Wakeley's morphological grades, height is a major criterion, with the lower
grades exhibiting greater seedling height. Similar results
have been published for loblolly and slash pine after 30
years (21).

Heights and diameters should both be considered when
developing predictions of field performance. If containerized seedlings are grown at high seedling densities,
heights may be about the same as when grown at lower densities, but stem diameters of the seedlings grown at the
lower densities will be larger--these perform better in the
field (3).

8.5 APPLICATION OF RESULTS

There is an insufficient amount of data at this time to
specify what the optimum characteristics of containerized
southern pine seedlings should be to obtain maximum survival
and growth when outplanted. Our best information is for
loblolly pine, but these data are from studies that were not
designed to provide predictive equations relating initial
seedling quality to field performance. However, these data
are probably the best available for the southern pines and
will give some feel for the relationship between seedling
physiology, morphology, and growth.

GROWTH

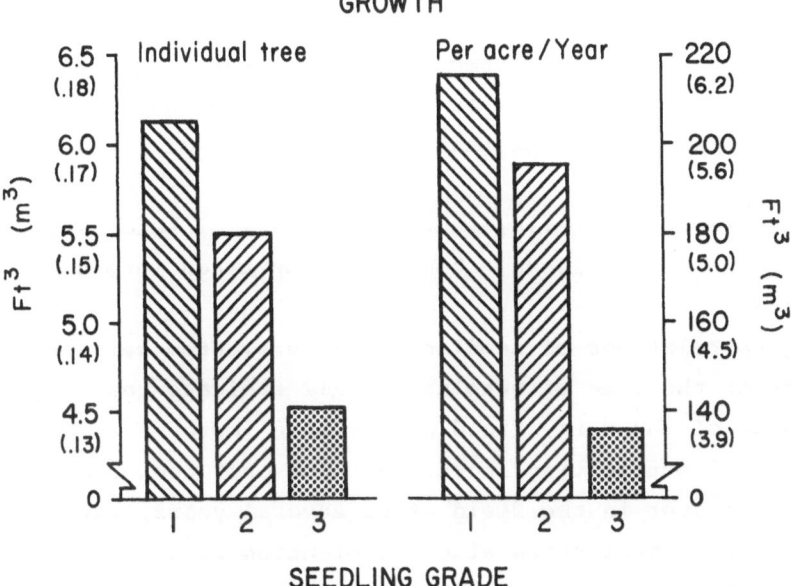

FIGURE 3. Volume performance of graded slash pine seedlings after 13 growing seasons (5).

There is considerable difference in the ease and reliability of measuring the various seedling characteristics that relate to field performance. Characteristics such as biochemical constituents, chlorophyll, and dry weights are not as easy to determine as are seedling heights and diameters. Our results indicate that some simplification of the types of measurements may be feasible. It must be realized, however, that such simplification will result in the loss of sensitivity that more elaborate techniques may provide. A number of the variables measured were directly related to initial seedling heights; for example, concentrations of lipids, sugars and chlorophyll, stem diameters, and stem weights. Although the data were not comprehensive enough to make all the correlations desired, several interrelations among seedling characteristics were shown, and a number of these could be directly related to field performance.

172

As long as the type of container and cultural treatment remain constant and provide good quality seedlings, height at the time of outplanting seems to be the best single indicator of field performance. It is easily measured and is related to other characteristics that are indicative of field success.

Other visual criteria, such as presence of secondary needles and woody tissue, should also be taken into consideration.

The correlations of seedling diameters, stem weights, and heights at the time of outplanting all indicate that, within the range of the data, field performance improves as size at the time of outplanting increases. Not only are seedling heights greater in the field after several years, but yearly growth for several years after outplanting is affected (FIGURE 4).

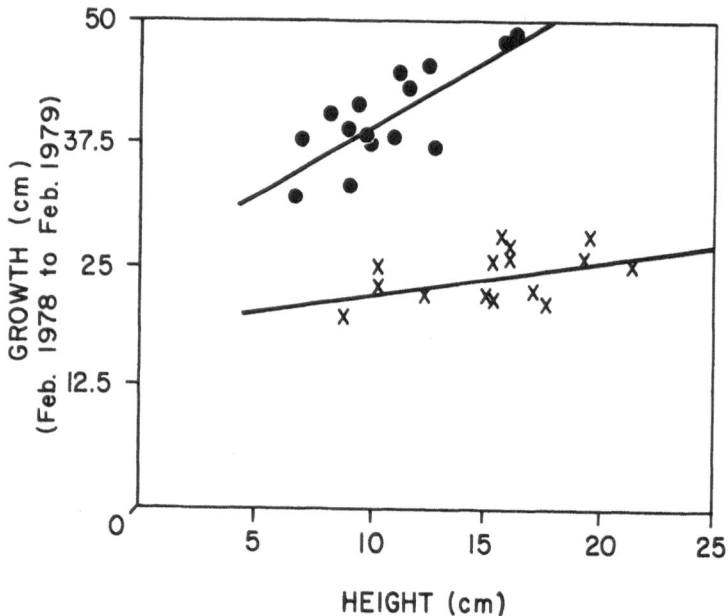

FIGURE 4. Relationship between initial height and growth over a 1-year period for loblolly pine seedlings (test 1) based on two outplantings.

This indicates that larger seedlings provide better field performance. However, these results reflect a limited range of initial seedling sizes--from about 8 cm to less than 23 cm in height. Also, it is expected that a point exists after which larger seedlings do not result in greater field growth. Data from another test, where initial seedling sizes were larger (18 to 33 cm), showed that there was no correlation with field heights in this larger size range (FIGURE 5). Thus, biologically as well as economically, there are practical limitations as to how large seedlings should be at outplanting.

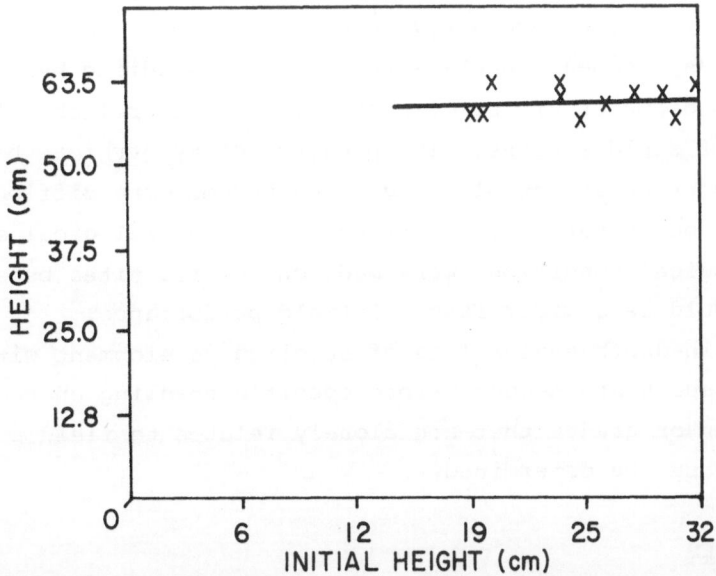

FIGURE 5. Initial height and field height relationships for loblolly pine seedlings 1 1/2 years after outplanting

Apparently morphological and physiological conditions do coincide, at least in some instances. At the present time morphological properties are easier to measure than physiological ones. There is a great need to continue to evaluate the physiological properties of seedlings associated with field performance and to develop simple and rapid measurement techniques. Particular attention should be given to parameters that were related to field success in

174

these tests: chlorophyll content, sugars, and lipids. Starch is another seedling constituent that may be indicative of seedling capacity to survive and grow.

8.6 CONCLUSIONS

At this time, there is insufficient data to reliably relate a wide range of physiological or morphological characteristics of container-grown seedlings to field performance, particularly survival. Several characteristics were related to initial seedling heights, which were closely correlated to growth in the field. But the relationships between physiological conditions and seedling performance have not yet been adequately researched.

The easy-to-measure characteristics of seedling height and stem diameter are still probably the most reliable indicators of field success. The presence of mycorrhizae becomes more important as the planting sites become more difficult. It would be better if evaluations of all physiological and morphological conditions were made on adverse sites because there would be a wider range of field performance.

More in-depth evaluations of seedling development with field results are needed before specific seedling characteristics or grades that are closely related to field performance can be determined.

REFERENCES

1. Barnett, J. P. 1974. Tube lengths, site treatments, and seedling ages affect survival and growth of containerized southern pine seedlings. U.S. Dep. Agric. For. Serv. Res. Note SO-174, 3 p. South. For. Exp. Stn., New Orleans, LA.
2. Barnett, J. P. 1980a. Containerized pine seedlings for difficult sites. In: Proc. Reforestation of disturbed sites, College Station, TX., June 9-11, 1980. Texas Agric. Extension Serv., Texas A&M Univ., [p. 51-58].
3. Barnett, J. P. 1980b. Density and age affect performance of containerized loblolly pine seedlings. U.S. Dep. Agric. For. Serv. Res. Note SO-256, 5 p. South. For. Exp. Stn., New Orleans, LA.

4. Barnett, J. P. 1983. Relating field performance of containerized longleaf and shortleaf pine seedlings to mycorrhizal inoculation and initial size. In: Proc. Seventh North American Forest Biology Workshop, Physiology and Genetics of Intensive Culture, Bart A. Thielges (ed.), University of Kentucky, Lexington, KY. July 26-28, 1982, p. 358-367.

5. Blair, R. and F. Cech. 1974. Morphological seedling grades compared after thirteen growing seasons. U.S. Dep. Agric. For. Serv., Tree Planters' Notes 25(1): 5-7.

6. Ching, T. M. 1963. Change of chemical reserves in germinating Douglas-fir seed. Forest Sci. 9: 226-231.

7. Davidson, W. H. and E. H. Sowa. 1974. Container-grown seedlings show potential for afforestation of Pennsylvania coal-mine spoils. U.S. Dep. Agric. For. Serv., Tree Planters' Notes 25(4): 6-9.

8. Ferdinand, I.S. 1972. Container planting program at North Western Pulp and Power Ltd. In: Proc. of a workshop on container planting in Canada. Environ. Canada, Canadian For. Serv., Dir. Program Coord. Inform. Rep. DPC-X-2, p. 21-25.

9. Iverson, R. D. and M. Newton. 1980. Large Douglas-fir seedlings perform best on Oregon coastal sites. Int. Pap. Co., Western Forest Res. Center Tech. Note 55, 9 p.

10. Jackson, M. L. 1958. Soil Chemical analyses. Prentice-Hall, Inc., Englewood Cliffs, N.J. 498 p.

11. Jorgensen, J. R. and E. Shoulders. 1967. Mycorrhizal root development vital to survival of slash pine nursery stock. U.S. Dep. Agric. For. Serv., Tree Planters' Notes 18(2): 7-11.

12. McGilvray, J. M. and J. P. Barnett. 1982. Relating seedling morphology to field performance of containerized southern pines. In: Proc., Southern Containerized Forest Tree Seedling Conference, Savannah, GA, Aug. 25-27, 1981. U.S. Dep. Agric. For. Serv., Gen. Tech. Rep. SO-37, p. 39-46. Southern Forest Exp. Station, New Orleans, LA.

13. Marx, D. H., J. L. Ruehle, D. S. Kenney, C. E. Cordell, J. W. Riffle, R. J. Molina, W. H. Pawuk, R. W. Tinus, and O. C. Goodwin. 1982. Commercial inoculum of *Pisolithus tinctorius* for ectomycorrhizal development of containerized tree seedlings. For. Sci. 28: 373-400.

14. Nelson, N. 1944. A photometric adoption of the Somogyi method for the determination of glucose. Jour. Biol. Chem. 153: 370-380.

15. Nalewaja, J. D. and L. H. Smith. 1963. Standard procedure for the quantitative determination of individual sugars and total soluble carbohydrate materials in plant extracts. Agron. Jour. 55: 523-525.

176

16. Ruehle, J. L., D. H. Marx, J. P. Barnett, and W. H. Pawuk. 1981. Survival and growth of container-grown and bare-root shortleaf pine seedlings with *Pisolithus* and *Thelephora* ectomycorrhizae. South. J. Applied For. 5: 20-24.

17. Shoulders, E. and J. R. Jorgensen. 1969. Mycorrhizae increases field survival of planted loblolly pine. U.S. Dep. Agric. For. Serv., Tree Planters' Notes 20(1): 14-17.

18. Sutton, R. F. 1980. Techniques for evaluating planting stock quality. International Union of Forestry Research Organizations (IUFRO Workshop, New Zealand August 1979). For. Chron. 56: 116-120.

19. Wakeley, P.C. 1948. Physiological grades of southern pines nursery stock. In: Proceedings, Society of American Foresters 1948: 311-312.

20. Wakeley, P. C. 1954. Planting the southern pines. U.S. Dep. Agric. For. Serv. Agric. Monog. 18, 233 p.

21. Wakeley, P. C. 1969. Results of southern pine planting experiments established in the middle Twenties. J. For. 67: 237-241.

22. Walker, N. R. and H. J. Johnson. 1980. Containerized conifer seedling field performance in Alberta and the Northwest Territories. Can. For. Serv., Northern Res. Centre Inform. Rep. NOR-X-218, 32 p.

23. Wood, J. P. and E. P. Bachelard. 1969. Variations in chlorophyll concentration in the foliage of radiata pine. Australian Forestry 33: 119-128.

Planting Site
and Stock Response

Matching Species and Stock Type to Site

Papers 9 and 10

9. THE INFLUENCE OF SPECIES AND STOCKTYPE SELECTION ON STAND
ESTABLISHMENT: AN ECOPHYSIOLOGICAL PERSPECTIVE

S. D. HOBBS

Associate Professor, Department of Forest Science, Oregon
State University, Corvallis 97331

ABSTRACT

The species and stocktypes planted can have a profound
impact on stand establishment. Selection of a particular
type of seedling able to meet minimum performance expectations
under specific types of environmental stress is only possible
after careful evaluation of site conditions likely to exist
at planting. These decisions have become less intuitive with
improved understanding of how seedlings respond physiologically to different operational environments. Our understanding, however, of the physiological bases for observed response
differences between stocktypes to the same site conditions
remains poor for many situations.

Physiological and morphological characteristics important
for adequate seedling performance under various types of environmental stress are discussed with emphasis on harsh-site
reforestation. A conceptual model for the selection of species and stocktypes best adapted to specific site conditions
is also presented.

9.1 INTRODUCTION

When the decision is made to forego natural regeneration
in favor of artificial methods, the silviculturist is faced
with a series of decisions that will influence stocking of
the site and stand growth. Two fundamental decisions that
must be made early in the planning process are which species
and stocktypes should be planted. The selections made may
ultimately affect the attainment of management goals. Decisions based on insufficient or misinterpreted information can

Duryea, M.L. and Brown, G.N. (eds.). Seedling physiology and reforestation succes
©1984, Martinus Nijhoff/Dr W. Junk Publishers, Dordrecht/Boston/London. ISBN 978-94-009-6139-5

produce disappointing results which may be apparent within
several years or not until later in the rotation.

The purpose of the ensuing discussion of species and stocktype selection is threefold. A framework is first provided
linking environment, physiology, and seedling response. This
is followed by an overview of current information regarding
species and stocktype selection, and subsequent seedling response to various types of environmental stress. Finally, a
discussion of methodologies for species and stocktype selection is given.

An assumption made throughout this chapter is that the
reader is aware of the importance of provenance selection.
Therefore, gains in seedling performance (survival and growth)
from genetic improvement, hybridization, and differentiation
between ecotypes are beyond the scope of this discussion. It
should also be recognized that genetic variation can be substantial even within a relatively small geographic area (90,
216) and that the application of genetic technology can have
a significant effect on reforestation success.

9.2 ENVIRONMENTAL INTERPRETATION

An important prerequisite to successful plantation establishment is the identification of those environmental factors
most likely to exert the strongest influence on seedling performance. However, foresters may fail to correctly assess
site conditions likely to exist at the time of planting.
This is largely a problem of perspective and how the individual views the environment and the changes that will affect
reforestation success.

9.2.1 The operational environment

If interpreted literally, the word "environment" is an
all-inclusive term used to define a plant's surroundings,
including both abiotic and biotic components. This seems
inadequate, however, in the functional sense if we accept the
premise that plant response is induced by internal physiological change and that these processes are in turn controlled by
the plant's genotype and external environment. Because a

plant's genotype is fixed for the life of that individual, environment is the dynamic influence. The suggestion that an interaction occurs between an organism and its environment led Mason and Langenheim (139) to propose the concept of "operational environment." This concept and that of "interaction" have been discussed in considerable detail by Spomer (188). Essentially the operational environment is that portion of the total environment that is composed of those factors capable of affecting an organism's physiology. Operational factors are real entities, either energy or material, that can be exchanged with the plant (Table 1).

Table 1. Generalized list of operational factors or factor groups (after 188; reproduced with permission from the Ecological Society of America).

Energy factors
    Ionizing radiation
    Light
    Heat
    Long wave
    Mechanical energy
    Miscellaneous energy (chemical, electrical, etc.)

Inorganic (mass) factors
    Gases
    Liquids and solutes
    Solids

Organic (mass) factors
    Nutrients
    Hormones
    "Communicators" (flower odors, etc.)
    Toxins and antibiotics
    Waste and by-products

In reforestation, operational factors that frequently limit success on difficult sites are heat, mechanical energy (animal damage), soil oxygen, and water. These are not the only operational factors that foresters need to be concerned with, but they represent the more common ones. The key is to identify those factors which are most limiting on a site-specific basis.

182

9.2.2 <u>Primary and secondary factors</u>

On forest sites one or more factors usually stand out as the most significant in terms influencing of plant response (10, 204). This concept is largely an outgrowth of Liebig's (128) "Law of the Minimum" and Blackman and Matthaei's (16) "Principle of Limiting Factors." Spurr and Barnes (189), however, note that Liebig's law is not always true because environmental factors frequently interact and that changes in a more abundant factor can compensate for a factor labeled as "limiting." Nonetheless, Hobbs (93) has suggested that on most sites scheduled for reforestation, a single operational factor can usually be identified as the potential primary cause of seedling mortality or growth reduction while other operational factors are of secondary importance. The rationale for this approach to site evaluation is that species and stocktype selection should be based on seedling characteristics that will offset or compensate for the potential primary cause of poor performance. Potential secondary factors can quite often be dealt with by ameliorative techniques or other protective measures.

Accurate site assessment in terms of the operational factors most likely to affect seedling performance is a skill developed as familiarity is gained with the ecosystems managed. In order for the appropriate species and stocktype to be available at the desired time of planting, site assessment must take place to allow adequate time for seedling procurement. Discussions of site evaluation are given by Cleary and Kelpsas (32), Daubenmire (43), and Greaves et al. (70).

9.2.3 <u>The ecophysiological perspective</u>

Seedling performance in the field is a function of changes in plant physiological processes which are influenced by the plant's genetic constitution and the surrounding operational environment. This functional relationship has importance to the forester both at the species and stocktype level. Since physiology is the integrator of genetic and environmental influences and subsequent plant response, an understanding of physiological change induced by a dynamic environment is

important. With intensive forest management we are now able
to improve the seedling's ability to withstand environmental
stress through genetic improvement, nursery cultural practices
and silvicultural manipulation of site and stand conditions.
A basic discussion of the ecophysiology of tree growth is
presented by Daniel et al. (42).

9.3 SPECIES SELECTION

Perhaps the most important reforestation decision that can
be made is the choice of species to be planted. It is not
inconceivable that in many cases, species selection has a
greater impact on future stand yield than any other decision
made during the reforestation process. On some sites species
selection may be relatively straight-forward. On others, it
can be complex, particularly for sites where several species
occur naturally. For example, in northeastern Florida and
southeastern Georgia, loblolly (<u>Pinus</u> <u>taeda</u> L.), longleaf
(<u>Pinus</u> <u>palustris</u> Mill.) and slash (<u>Pinus</u> <u>elliottii</u> Engelm.)
pines frequently occur on the same imperfectly drained flat-
woods sites. Although longleaf pine was once the dominant
species on many of these sites, it has been shown that slash
pine produced more volume after 26 years (157).

9.3.1 <u>Ecological amplitude</u>

The natural range of any species is limited by its ability
to successfully reproduce (204), which is a function of its
ecological amplitude or silvical characteristics. Daubenmire
(44) has defined this as: "The breadth of environmental
requirements and tolerances of a taxon." If viewed within
the context of the operational environment, then there are
maxima and minima for each component of the operational environment within which a given species can successfully complete
its life cycle. As these limits are approached, first growth
and then the ability to reproduce are diminished. The fact
that species respond differently to various forms of stress
in the operational environment, has been shown in a large
number of ecophysiological studies (24, 36, 67, 98, 130, 220).
Descriptions of the natural range or habitat requirements of

most native tree species in the United States are readily
available (62). In the Pacific Northwest Minore (148) has
also compared the autecological characteristics of many
Northwest tree species.

9.3.2 Standard guidelines

   Planting only indigenous species probably best describes
the general philosophy for species selection practiced today.
Anderson (4 quoted in 187) states that above all else, the
species chosen should be adapted to the site environment.
Once the adaptational requirement has been satisfied, "the
species selected from those suited to the site should be the
ones that promise the best net returns" (187). Other, more
recent discussions of species selection in the United States
are also available (106, 164, 183, 186, 214). A comprehensive
examination of species selection on a world-wide basis has
been presented by Champion and Brasnett (27).

9.3.3 Monocultures and mixed stands

   There are probably few objective, scientifically based
guidelines for deciding whether to establish a monoculture or
a mixed stand. Foresters should, however, recognize the necessity of evaluating each situation independently. The decision
is most complex on those sites where more than one economically
important species is well-adapted to the operational environment. The problem is epitomized on the more productive sites
where both conifers and hardwoods are commercially important.
In such cases the decision is largely based on an economic
evaluation; except in those cases where other resource values
are of prime importance such as providing adequate wildlife
habitat. One of the important considerations in any analysis
of economic alternatives is the potential for loss from disease
and insect pests or other natural disasters. Historically,
stands of mixed species composition are more resistant to
disease and insect damage than monocultures (19, 69 cited in
187). This, of course, is a generality to which there are
numerous exceptions (161). For example, there is evidence to
indicate that vigor is important to tree resistance to insect

attack as indicated by the fact that vigorous monocultures of lodgepole pine (<u>Pinus contorta</u> Dougl.) and ponderosa pine (<u>Pinus ponderosa</u> Laws.) are less susceptible to attack by mountain pine beetle (<u>Dendroctonus ponderosa</u> Hopk.) (122, 215). For certain pests, however, such as dwarf mistletoe (<u>Arceuthobium</u> spp.), increased tree vigor generally means increased susceptibility to infection (162).

Discussions of the relative advantages and disadvantages of monocultures and mixed stands are contained in Hocker (97) and Smith (187). The world-wide perspective on stands of mixed species composition is again given by Champion and Brasnett (27) while information on specific forest types within the United States may be found in Barrett (12) and U.S. Department of Agriculture (209).

9.3.4 <u>Introduced species</u>

Experimentation in the United States with exotic species has not resulted in wide-spread operational acceptance of any species on a scale similar to that of Monterey pine (<u>Pinus radiata</u> D. Don) in New Zealand or Douglas-fir [<u>Pseudotsuga menziesii</u> (Mirb.) Franco] in Europe. A potential exception is the use of eucalypts in some areas of the United States. Although limited primarily by freezing temperatures, eucalypts have been successfully established in parts of California, Florida, and Hawaii (102). Several species have the potential for substantial volume production in Florida (100). The use of English oak (<u>Quercus robur</u> L.) may also have potential in some parts of North America (104, 219). Attempts to extend the natural range of native tree species have not been particularly successful from the operational perspective either. Extension southward of red pine (<u>Pinus resinosa</u> Ait.) and the movement of loblolly pine, slash pine and redwood [<u>Sequoia sempervirens</u> (D. Don) Endl.] northward have all been largely unsuccessful (105 cited in 97, 187, 195). It should be noted, however, that loblolly pine has been grown successfully in southern Illinois (66) although some seed sources are sensitive to extreme cold weather (65).

Whether extending the natural range of native species or
introducing species from other geographic regions, an extended
period of experimentation should be undertaken before large
operational plantings are attempted. In general, introduced
species should only be considered if the operational environment into which they will be placed approximates that of their
natural range, and the potential economic benefits to be gained
exceed those of native species. A more detailed discussion
of this subject is provided by Wright (218).

9.3.5 Some final remarks

The process of species selection is one that requires knowledge of the silvical characteristics of the species, understanding of the operational environment into which it will be
placed, and the likely response of the individual tree or
stand. The biological consideration is not, however, the
only major factor influencing the choice of species to be
planted. Obviously, the management or return on investment
objectives are of concern and offer a framework within which
a reforestation prescription may be written.

9.4 STOCKTYPE SELECTION

The process of matching seedlings to the planting environment should not end once the choice of species has been made.
Unfortunately, the effect that seedling stocktype can have on
subsequent field performance may often be overlooked. Surely
many plantation failures could be avoided if foresters were
aware of the various stocktypes available to them and how
each would respond to a particular operational environment.
The fact that several stocktypes may respond differently in
the same operational environment is well-documented by a large
number of studies (3, 8, 50, 74, 94, 95, 152, 159, 174 and
others). Unfortunately, few studies have examined the physiological bases for these response differences. Rather, the
majority of stocktype comparisons have concentrated on measurements of survival and growth. There are two notable exceptions, however. These are studies of root wrenching and seedling inoculation with mycorrhizal-forming fungi. Duryea and

Lavender (54) and Walker et al. (213) are suggested as introductory references to the literature regarding these topics.

9.4.1 Defining stocktype

The term "stocktype" is typically used to separate seedlings into distinct categories based on their age, size, and
the cultural method of rearing (e.g., bareroot, containerized,
etc.). In recent years the ability to alter both the seedling's morphological and physiological characteristics by
changing or modifying cultural techniques, has greatly increased. This has resulted in an ever-expanding array of
stocktypes. To cope with this, Nicholson (155 cited in 31)
developed a standardized descriptive code that incorporates
seedling height, caliper, and shoot-root ratio. This code
may be expanded to include the type of seedling (e.g., bareroot, containerized) and the proposed planting season (31).
Cleary et al. (31) stresses that the code is simply a communi-
cative device and is not necessarily indicative of physiological vigor.

9.4.2 Types of stock

9.4.2.1 Bareroot. Bareroot seedlings are widely planted
throughout North America and are generally from one to three
years old when lifted. In 1981, 253 nurseries produced bareroot stock in the United States (6). A comprehensive report
on the biology of bareroot seedlings, their response to various cultural practices and the management of bareroot nurseries has been presented by Duryea and Landis (53). Discussions
of bareroot nursery operations are also contained in other
publications (1, 7, 31, 183, 210).

Bareroot stock are reared from seed sown in the nursery
bed and left in place until lifted. Desired morphological
and physiological characteristics can be manipulated by modification of the seedbed density, regulation of the irrigation
regime, fertilization, undercutting and wrenching, lateral
root pruning, inoculation with mycorrhizal fungi, and lifting
date. One- and two-year-old bareroot seedlings are most frequently used. In the Pacific Northwest, 2-0 stock are planted

on a wide variety of sites but 1-0 seedlings are rarely used because foresters thought that they lacked adequate root systems, caliper, and height for many site conditions. This assumption, however, may be erroneous. For example, in a recent experiment on a droughty, south-facing clearcut in southwest Oregon, I found that 1-0 bareroot ponderosa pine seedlings survived and grew in height at rates similar to those of 2-0 stock after two years of observation (Table 2). In the southeastern states, 1-0 pine seedlings are routinely planted in many areas because under such favorable nursery growing conditions, they are of ideal planting size after just one year. One-year-old bareroot seedlings are also less expensive to produce. Early results in northern California and southern Oregon with 1-0 Douglas-fir seedlings have also been encouraging (103).

Table 2. Total percent survival and mean annual height and diameter growth (±SE) of ponderosa pine 1-0 and 2-0 bareroot stocktypes on a droughty site in southwest Oregon.

Bareroot stocktype	Survival (%)		Height growth (cm)		Diameter growth (mm)	
	1981	1982	1981	1982	1981	1982
1-0	99	99	4.0±0.2	13.2±0.4	1.8±0.1	6.3±0.2
2-0	98	97	5.1±0.5	13.5±0.7	1.6±0.1	6.4±0.2

Transplants are also bareroot seedlings but have been lifted from one growing environment and placed in another prior to outplanting. Transplanted seedlings are typically larger than 1-0 and 2-0 bareroot seedlings or container-grown stock. Seedlings that have been transplanted generally fall into two categories depending upon the initial cultural method used in their rearing. They are either raised first in containers and then transplanted into the nursery bed (e.g., plug-1 stock) or grown in the nursery bed and then transplanted into another bed at a lower density (e.g., 1-1, 1-2, 2-1, 2-2 stock, etc.). The plug-1 seedling is a relatively new stocktype for which little field performance data is available, although several comparative studies have been reported for Oregon (74, 95). Therefore, the appropriate utilization of plug-1 seedlings has not yet been clearly iden- tified. Information on the field performance of the more traditional types of transplants, compared with other stock- types, is more readily available (2, 55, 57, 89, 101, 152).

9.4.2.2 Container. During the last 15 years container- grown seedlings have gained in popularity. In 1981 there were approximately 92 nurseries in the United Stated produc- ing various types of container stock (6). In general, container-grown seedlings are smaller and younger than bare- root stock (191). There are two basic types of containers: those from which the seedling and potting mix are extracted just prior to planting (plugs), and those that are planted while the seedling is still in the container (bullets) (202). Most bullet containers are either designed to biodegrade or break apart as the roots egress.

Container-grown stock offer a number of advantages. They are usually of good quality because of the controlled envi- ronment in which they are grown; the root systems are not disturbed until planting which eliminates or minimizes dan- gerous root exposure; the planting season can be extended; production lends itself to mechanization; and they are easier to plant, particularly in rocky soils. However, container- grown stock of some species, such as Douglas-fir, may not perform as well as bareroot stock in certain environments such as cool, moist coastal sites (8, 101). Container seed- lings are also expensive to produce although a recent study in the southern United States showed that they were no more costly to produce and plant than bareroot stock (71).

A thorough description of container technology has been provided by Tinus and McDonald (202). Other more general discussions are given by Cleary et al. (31), Hahn (73), Kinghorn (107), and Stein (190). Major conference proceedings dealing with container technology in North America (203), the southern United States (72), and Canada (178) have also been published. Several papers on root form and associated prob- lems in container stock are also discussed in a proceedings edited by Van Eerden and Kinghorn (211).

9.4.2.3 <u>Wildlings</u>. Wildlings are seedlings that occur naturally and are lifted from areas near the planting site. They are rarely used in operational, large-scale reforestation efforts because finding and lifting an adequate number of trees is difficult and expensive. Wildling root systems may also frequently be inadequate for transplant purposes. Wildlings may, however, be useful in cases where only a few trees (perhaps several hundred) are needed to restock a small area and better seedlings are unavailable.

Newton (154) reported on the performance of transplanted western hemlock [<u>Tsuga heterophylla</u> (Raf.) Sarg.] wildlings in the coast range of Oregon. After four years wildling survival was 66 percent in one treatment with reasonably good height growth. Unfortunately, other stocktypes were not planted at the same time and comparative data are not available. In another study, however, using red alder (<u>Alnus rubra</u> Bong.), transplanted wildlings survived poorly when compared with container-grown seedlings (83).

Other information regarding this stocktype is almost totally absent from the literature. This probably reflects the general opinion, among silviculturists, that wildlings do not compare favorably with other stocktypes.

9.4.2.4 <u>Cuttings</u>. Vegetative propagation, using stems cut from parent trees, has been used for many years to establish plantations and perpetuate desirable genotypes in genetic research and tree improvement programs. The operational use of cuttings is, however, limited to a relatively small number of species, sites, and management objectives. For example, this technique has been more successful with certain hardwood genera than with conifers, although many conifer species have been successfully rooted.

Hardwood cuttings are widely used in the southern states where the long growing season promotes rapid growth, particularly on sites where moisture is not limiting. Cottonwood species (<u>Populus</u> spp.) in particular have been successfully established from cuttings both in the South (143) and in Washington (153). However, for some species, such as American sycamore (<u>Platanus occidentalis</u> L.), bareroot seedlings

do better than cuttings (119) and are normally recommended
for plantation establishment (21). Other references dealing
with vegetative propagation with cuttings are Briscoe (21),
Hansen and Phipps (75), Hansen and Tolsted (76), Hare and
Land (77), Heilman and Ekuan (82), Krinard (117), Miller et
al. (145), and Wright (218)

9.4.2.5 Germinants. Little research has been reported on
germinant reforestation except that of Anderson and Williamson
(5) in California and later work done by DeVelice and Buchanan
(46) in New Mexico. Anderson and Williamson found that Coul-
ter pine (Pinus coulteri D. Don) seeds germinated first on
vermiculite-filled flats and covered with moist burlap, could
be successfully field-transplanted in southern California
when the radicals were between 13 and 38 mm in length. When
these germinants were covered with screen caps to protect
them from rodents and birds, their survival rate was much
better than that of 1-0 bareroot seedlings planted at the
same time. DeVelice and Buchanan found similar results with
ponderosa pine germinants in the Sacramento Mountains of New
Mexico, although they did not have side-by-side trials with
other stocktypes.

Despite the successes discussed by Anderson and Williamson
(5) and DeVelice and Buchanan (46), no further work with conifer germinants has been reported. The reason for this apparent lack of interest is not clear, although this seems an
area where future research may be justified.

9.4.2.6 Mycorrhizal stock. In the last ten years, our
understanding of mycorrhizae has made it possible to successfully inoculate seedlings in the nursery with specific
mycorrhizal-forming fungi and thus improve seedling performance after planting (37, 174). Seedlings that have been
successfully inoculated with mycorrhizal fungi may experience
morphological or physiological changes that enhance their
ability to cope with various types of environmental stress.
For this reason, seedlings having mycorrhizae as the result
of inoculation, should also be considered as a distinct stocktype. The benefits and importance of symbiotic fungi that
form mycorrhizae with coniferous and hardwood tree seedlings

are well-documented and general discussions have been presented by Marx (133), Marx and Barnett (134), Mikola (144), Molina
(149), and Trappe and Fogel (206). Mycorrhizae are generally
classified into two groups: endomycorrhizae and ectomycorrhizae. These groups are separated by their infection anatomy
(133, 206) although Trappe and Fogel (206) have stressed that
endomycorrhizae might be better understood if divided into
three separate groups. Most research, however, has concentrated on the ectomycorhizal fungi associated with coniferous
species although in recent years increasing interest in endomycorrhizae and mycorrhizae on hardwoods has been evident. A
detailed review of ectomycorrhizae ecology and physiology has
been presented by Marks and Kozlowski (132), but Lamont (118)
has given a more recent account of nutrient uptake in plants
as affected by both ecto- and endomycorrhizae.

There seems little doubt that the use of host- and site-
specific ectomycorrhizal fungi on an operational scale, in
both container and bareroot nurseries, is feasible in the
very near future as evidenced by recent publications (37,
120, 138, 205). That mycorrhizae enchance seedling performance under a wide variety of operational environments is
obvious from the large body of literature devoted to reports
of specific studies (9, 41, 49, 50, 63, 109, 136, 174, 213).

9.4.3 Problem sites

Foresters are often faced with the problem of reforesting
areas where the achievement of a minimum stocking level in a
specific period of time is doubtful because of some inherent
site problem related to the operational environment. Typically, the margin for error on these sites is smaller than
that associated with more moderate environments. Consequently, careful planning and execution of the reforestation
process carries added significance. In cases where there is
flexibility in the choice of species, the effects of a specific operational factor may be changed by selection of the
species best-adapted to the specific problem. Once this
decision has been made, further improvements in performance
can often be realized by use of a stocktype specifically

matched to site conditions. Where flexibility in the choice
of species does not exist, stocktype selection may take-on
added importance.

9.4.3.1 <u>Drought</u>.[1] The influence of decreased available
moisture on tree response has probably been investigated more
than any other operational factor, particularly in conditions
of increasing evaporative demand. This is understandable
considering that water is involved in all physiological processes (39) and is probably the most important environmental
factor limiting tree distribution and growth (25).

Brix (22) found that the rate of photosynthesis in loblolly
pine seedlings declined rapidly at xylem pressure potentials
less than -6 bars while in Douglas-fir seedlings this point
was reached at -10 bars (28). The effects of water deficits
on photosynthesis, and the probable mechanisms by which this
takes place, have been recently reviewed (20). Prolonged
periods of moisture stress can cause the cessation of cambial
activity (113, 221), inhibit cell division (114), and result
in smaller leaf size (111). Moisture stress can also decrease
carbohydrate translocation (169) and soil drought can reduce
root growth (13, 126, 156, 168, 193, 222) as well as nutrient
uptake, although to a lesser degree (118). Kozlowski (110)
points out that water deficits can predispose plants to attack
by certain pathogens and insects as well.

Numerous studies have compared the performance of various
stocktypes on droughty sites but few have examined the physiological reasons for differences in survival and growth, except
as previously noted (Section 9.4) for mycorrhizal and wrenched
seedlings. Furthermore, no clear consensus favoring a particular stocktype is evident after a review of the literature.
This may be partially attributable to environmental variation
between test sites and unequal seedling quality between the

---

[1]The term "drought" is used in this discussion to mean low
available soil moisture as opposed to "physiological drought"
(31) which also includes flooding and low soil temperature as
causal factors.

stocktypes compared. The latter poses a rather provocative question that centers on the issue of whether or not stock quality is correlated with the nursery cultural method used (e.g., container versus bareroot). Comparative tests of phy- siological vigor or seedling quality as described by Ritchie (167), have also not been conducted between stocktypes.

One such test is the measurement of root growth potential (RGP), an important characteristic for seedlings to be planted on droughty sites. Much has been written about RGP (also referred to as root regeneration potential and root growth capacity) over the last 30 years, most of which has been sum- marized by Ritchie and Dunlap (168). There seems little doubt that RGP can be used as a measure of physiological vigor as demonstrated with ponderosa pine in California (192, 194, 196, 197). Ritchie and Dunlap (168, citing numerous studies) also point out that the period of maximum RGP coincides with the time when seedlings are most tolerant of desiccation and handling damage. Using bareroot lodgepole pine seedlings, Burdett (26) observed a correlation between RGP and seedling mortality, thus reinforcing its importance to seedling survival.

In environments where conditions favorable for growth exist for only a few months before moisture is limiting, it is essen- tial that newly planted seedlings extend their roots in order to maintain contact with receding moisture (182). Intuitively, it is reasonable to suggest that on droughty sites, the single most important seedling characteristic that will influence survival and growth during the first year is probably RGP. This premise must be tempered, however, by the knowledge that dormancy intensity at lifting is also important to physiologi- cal vigor and field performance (91, 123, 125). Evidence summarized by Ritchie and Dunlap (168) suggests that RGP and dormancy intensity are related, although the mechanism by which this occurs is obscure.

When choosing a stocktype to be planted on a droughty site, a logical question to ask is whether or not there are inherent differences in RGP between stocktypes; particularly those grown by dissimilar cultural techniques (e.g., container versus bareroot). If differences exist, are they consistent from year-to-year or simply chance occurrences. Unfortunately, this question is unresolved because RGP comparisons between stocktypes have not been reported as previously mentioned. In a study of three Douglas-fir stocktypes in southwest Oregon, however, Hobbs and Wearstler (95) did find large differences in predawn xylem pressure potential between stocktypes (Figure 1) and subsequent field performance. Although RGP was not quantified, differences in new root growth were striking when sample seedlings were excavated at three-week intervals during spring and summer. Container-grown 1-0 plug seedlings had the highest survival, least negative xylem pressure potential, and greatest root growth when compared to 2-0 and plug-1 bare-root stock.

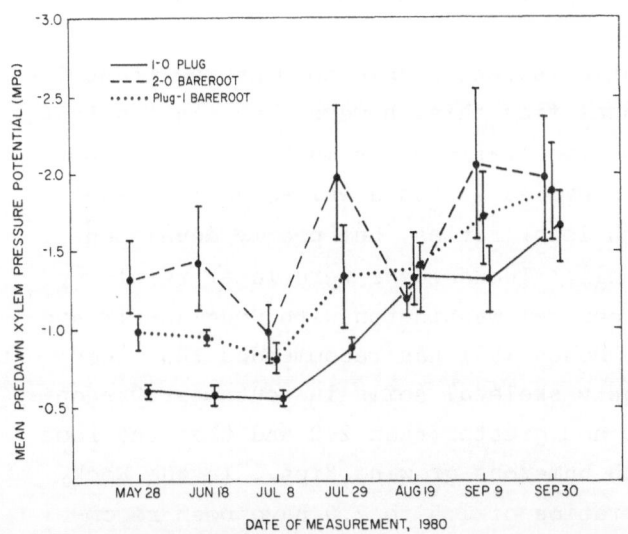

FIGURE 1. Mean predawn xylem pressure potential (±SE) of three Douglas-fir stocktypes planted on a droughty site in southwest Oregon (after 95).

In stocktype comparison studies in the western United States and British Columbia, container-grown stock have been shown to perform better than bareroot stock on a variety of droughty sites (8, 74, 94, 95, 142, 159). This same trend was also reported in the southeastern United States for a study conducted during an exceptionally dry year (3). In another study, leaf water potential in bareroot black oak

(Quercus velutina Lam.) seedlings was more negative than in container-grown stock during a mild drought in Missouri (50).

In spite of these reports, there is still reason to question the wholesale prescription of container-grown stock for all droughty sites. For example, Helgerson (85) found that 2-0 bareroot Douglas-fir and ponderosa pine seedlings survived at significantly higher rates than 1-0 plug stock after two years on a droughty site in southwest Oregon. Other studies, although not comparing bareroot and container seedlings, reinforce Helgerson's findings, since they show that good quality bareroot seedlings can survive at high percentages on a variety of droughty sites (103, 129, 199, 200).

Hermann (88) suggested that on droughty sites, seedling shoot-root ratio is an important morphological characteristic, explaining that the root system of a seedling with a more balanced shoot-root ratio is better able to meet the transpirational demands. He qualifies this, however, by stating that a seedling with a shoot-root ratio of as much as 3.0 may not be adversely affected provided it has a well-developed root system. Seedlings with lower ratios, but poorly developed roots, may not do as well. These ideas were later reinforced by work done in north-central Washington with Douglas-fir and ponderosa pine (129). Hobbs (93) has recommended that seedlings planted on droughty skeletal soils in southwest Oregon have shoot-root ratios not greater than 2.0 and that the root systems be fibrous with numerous growing tips. In the Rocky Mountains, shoot-root ratios of 1.5 to 2.0 have been recommended for ponderosa pine (61).

Other studies in which shoot-root ratios have been decreased as a result of specific nursery cultural practices (undercutting and wrenching) indirectly support the idea that smaller shoot-root ratios are beneficial on droughty sites (58, 108, 172, 201). It should be noted, however, that this is an association and that no cause-and-effect relationship has been demonstrated. This is particularly true since other studies of undercutting and wrenching have shown that seedling response to droughty environments does not improve despite decreased shoot-root ratios (54, 96).

Cleary et al. (31) have stated that "shoot-root ratios should not be used as a single indicator of seedling field survival potential." Recently, in a large quantitative study, shoot-root ratio was shown to be poorly correlated with field performance and was found to be a poor indicator of stock quality, although this study was not conducted on droughty sites (165). Recognizing the necessity for some degree of balance between shoots and roots, Ritchie (167) correctly points out that traditional methods of measuring shoot-root ratio (weight and volume) do not really measure the root system's ability to provide water and minerals. A more appropriate measure would be one of absorption capacity, which unfortunately is difficult to measure.

Evidence from field trials on droughty sites indicates that seedlings successfully inoculated with mycorrhizal-forming fungi perform better than uninoculated seedlings. In experiments with white oak (Quercus alba L.) and black oak seedlings inoculated with the fungal symbiont Pisolithus tinctorius, (Pers.) Coker and Crouch, Dixon et al. (50, 51) demonstrated that mycorrhizal seedlings have improved drought resistance. Parke et al. (160), using Douglas-fir seedlings grown under greenhouse conditions, found that drought-stressed mycorrhizal seedlings fixed CO_2 at a rate ten times greater than that of non-mycorrhizal seedlings. In eastern Washington, however, Bledsoe et al. (17) found that 17 months after outplanting, mycorrhizal container-grown Douglas-fir seedlings did not differ significantly from non-mycorrhizal seedlings in either survival or height growth. This may in part have been due to the fact that new roots of all seedlings became colonized by local mycorrhizal-forming fungi within five months. The authors did observe that the inoculated fungal symbionts did not appear to compete well with the indigenous species. This emphasizes the importance of matching the inoculum source as closely as possible to planting site conditions, a point also made by Trappe (205). A recent review of the potential impact of mycorrhizae on seedling performance in a droughty environment has been provided by Molina (150).

Three general characteristics that will, in all likelihood, improve seedling response on droughty sites are:

- high root growth potential.
- fibrous root systems with numerous growing tips.
- shoot-root ratios of 2.0 or less. (This still seems prudent despite conflicting opinions. A balanced shoot-root ratio may be of little consequence, however, if other characteristics are lacking.)

9.4.3.2 Flooding. The effects of flooding on woody plants has been discussed in detail in several publications (112, 116). Soil inundation results in anoxic or oxygen-deficient conditions that cause reduced water absorption (this is another form of physiological drought), decreased mineral uptake, and leaf abscission (112), as well as a variety of other problems. Kozlowski (112 citing 64 and 173) states that "The extent of injury by flooding varies by species, soil factors, timing and duration of flooding, and physico-chemical conditions of the floodwater." Coutts and Armstrong (38) suggest that plant age and dormacy status are also important to the damage done to flooded plant roots.

Obviously many tree species are native to habitats where periods of flooding or high water tables are common occur-rences of varying duration. The adaptations that many of these species have to withstand anaerobic soil conditions are presented by Coutts and Armstrong (38) and Crawford (40). Other discussions of the effects of flooding on roots can be found in Orlov (156) and Troughton (207). Almost all previous work devoted to the problem has been conducted at the species level examining tolerance and response differences (34, 98, 99, 141, 146, 147, 220). An exception to this is in the southern United States where the use of seedlings rather than cuttings is recommended for the establishment of cottonwood (Populus spp.) plantations in areas subject to flooding (131). It is thought that root development during the first year is important to survival (117).

9.4.3.3 <u>High temperatures</u>. Heat directly affects seedling growth (33, 87, 184) because "it influences many physiological activities through its affect on rates of metabolism" (25). The direct effects on physiological processes, such as photo-synthesis, are not easily quantified, however, because of interactions with other factors such as moisture (Helms 86). Other discussions of the effects of heat on tree growth and physiology are provided by Kramer and Kozlowski (116), Lavender (124), and Orlov (156).

Heat damage to planted seedlings is difficult to identify because it is often associated with high levels of moisture stress. Greaves et al. (70) have suggested that soil surface temperatures above 54°C will cause physical damage to seedlings less than six weeks old. Shirley (185 cited in 116) found that older and larger seedlings were more resistant to heat damage. It has been suggested that seedlings of larger stem diameter are able to dissipate heat and are better insulated by the corky cortex (29, 31). Damage in the form of heat lesions can also occur on older seedlings. For example, I have observed mortality in 2-0 bareroot Douglas-fir seedlings from heat lesions formed at the soil line in southern Oregon. This is a form of girdling which can be lethal if the lesion covers the total circumference of the seedling stem. Recent investigations have shown, however, that shadecards can improve Douglas-fir seedling survival on hot south aspects (92, 163), although a cause and effect relationship between lower soil surface temperature and survival was not examined.

In controlled environments, mycorrhizae have been shown to improve seedling performance when subjected to high root-zone temperatures. Experimenting with loblolly pine seedlings, Marx et al. (137) were able to demonstrate improved seedling vigor at root-zone temperatures of 29°C with an isolate of <u>Pisolithus</u> <u>tinctorius</u>. Seedling vigor did decline, however, at 34°C even though mycorrhizal development increased, sug-gesting that a physiological limitation in the host had been reached. The following year Marx and Bryan (135) reported on additional research done with loblolly pine seedlings inocu-lated with mycorrhizal-forming fungi and grown at high root-zone temperatures. Seedlings inoculated with <u>Pisolithus</u> <u>tinctorius</u> and <u>Thelephora</u> <u>terrestris</u> (Ehrh.) Fr. survived better (95 and 70 percent, respectively) than nonmycorrhizal seedlings (45 percent) after five weeks at a root-zone temperature of 40°C. They concluded that ectomycorrhizal seedlings that had been inoculated with the fungal symbiont <u>Pisolithus</u> <u>tinctorius</u> should be able to withstand high soil temperatures better than nonmycorrhizal seedlings or seedlings with <u>Thelephora</u> <u>terrestris</u>. The exact physiological reason for improved tolerance of high root-zone temperatures by mycorrhizal seedlings is not known although it seems reasonable to speculate that improved internal water potential may be a contributing factor.

Cases in which planted seedling mortality can be directly linked to high heat are probably rare except for unusual sites such as those represented by some mine spoils. As previously mentioned, separation of heat damage from that caused by high internal water deficits or their collective influence is difficult, and for this reason stocktype recommendations given for droughty sites (Section 9.4.3.1) probably have equal applicability. The importance of mycorrhizae to heat tolerance of seedlings has been demonstrated, but specific recommendations should only be made after careful localized experimentation has been conducted to evaluate host-fungus-site interactions. Large-diameter seedlings should also be used, costs permitting, provided the stock recommendations given in Section 9.4.3.1 can be met. This is particularly important since moisture stress may decrease the seedling's ability to withstand high temperatures.

9.4.3.4 <u>Low temperature</u>.[2] Damage from low temperatures can occur at any time of the year but is most serious when sudden frosts develop during the growing season (19). Upon freezing, ice crystals form in the intercellular spaces dehydrating the protoplasm (68). Ice crystals may also form

---

[2]Temperatures at or below 0°C.

within cells disrupting the cytoplasm (179 cited in 68) although this type of damage is considered rare in nature (177,
179). Frost-sensitive plants are unable to adequately rehydrate and resume normal metabolism after thawing, whereas the
cells of hardy plants are able to rehydrate as the ice crystals melt (177). Even though most north-temperate tree species naturally possess some degree of frost hardiness, substantial frost damage can occur if cultural regimes have not
been initiated in the nursery to induce the hardening process
well before lifting. Unfortunately, frost damage that has
occurred in the nursery may not be noticed until after planting. Even then it may not significantly affect seedling performance. For example, Edgren (56) found that Douglas-fir
seedlings damaged by an early frost in September while still
in the nursery bed were unaffected in terms of survival and
height growth three years after planting. On the other hand,
I have observed very high mortality in frost-damaged 2-0 bareroot and 2-1 transplant Douglas-fir seedlings within a few
months of planting. Undoubtedly, there are several factors
that interact to influence the expression of frost damage in
seedlings (i.e., frost intensity, degree of hardiness at the
time of frosting, outplanting site environment, etc.).

Frost hardiness is an important and complex seedling at-
tribute, the physiological basis for which is poorly understood. Its importance to seedling survival on sites where
freezing temperatures occur is well-documented and several
excellent discussions of the subject are suggested as introductory references to a voluminous literature (68, 116, 127).

Research on the frost resistance of forest tree seedlings
has concentrated on species differences (36, 217). The latter
and Glerum (68) have suggested that breeding programs should
select for frost resistance. There are, however, no specific
stocktype guidelines that I am aware of that have shown stocktype to be a significant factor in resistance to frost damage.

Low soil temperatures can cause lethal water deficits to
develop in seedlings when high evaporative demands are placed
on exposed tops (another from of physiological drought) while
the roots are below melting snow (175). As soil temperatures

approach 0°C, water viscosity increases while water flow and
root permeability decrease (175). Damage that results from
this condition is often referred to as "winter desiccation."
On those sites where the potential for serious winter desiccation exists, fall planting should probably be avoided and
only seedlings with relatively balanced shoot-root ratios and
fibrous root systems used. Intuitively, it also seems reasonable to suggest that high RGP would be another desirable seedling characteristic, although the benefit of increased root
absorptive area might not be realized until the following
year. These recommendations are based largely on supposition
and are not supported by data documented in studies of winter
desiccation since little effort has been devoted to this particular problem.

Another important cause of seedling mortality associated
with low temperatures is frost heaving (80). Heidmann (80
citing 181 and others) states that frost heaving of seedlings
primarily involves the soil surface layer and that this process can be separated into three phases (81). First, the
seedling stem is held tightly as the surface layer of soil
freezes. This is followed by the formation of ice lenses
lower in the soil profile which lift the frozen soil and
seedling. As the soil thaws and settles, the seedling is
left on the surface. For a more detailed discussion see
Cochran (35) and Heidmann (80).

Seedlings that have been subjected to frost heaving will
in all likelihood die depending upon the extent of root exposure. For example, Larson (121) reported 52 percent mortality of ponderosa pine seedlings after a single frost heaving
event. Frost heaving appears to be most serious in young
seedlings less than one-year-old, and is less severe in 2-0
and larger stock (80). As Heidmann (80) points out, there
may be two factors that explain the apparent resistance of
older seedlings to frost heaving. Larger seedlings offer
more wind resistance which causes them to sway back-and-forth
creating a gap between the seedling stem and the surrounding
wet soil which prevents the tight gripping of the stem. When
the soil water freezes, the seedling is left in place as

lifting of the surrounding soil occurs (181 cited in 80). Older seedlings also generally have larger root systems which better anchor the seedling against heaving (80).

Many foresters believe that container-grown seedlings are more susceptible to frost heaving than bareroot stock. I have found little documented evidence to indicate that this is a wide spread problem, although there have been instances when this has occurred. For example, Heidmann and Thorud (81) found that ponderosa pine seedlings in Ontario tubes were uplifted much more than plugs which were hardly moved by frost heaving. Severe frost heaving of Ontario tubes has also been reported for mine spoil sites in Pennsylvania (45). Cleary et al. (31) have suggested that frost heaving of container-grown stock may be particularly serious when dibble-planted in soils that are easily glazed. If container-grown seedlings are used it is recommended that fall planting be avoided whenever possible in areas prone to frost heaving because the seedlings may not develop adequate roots to anchor themselves before frost heaving begins. In this regard, RGP would seem an important characteristic. Care should also be taken to ensure that seedlings are not planted shallow or dibbles used as the planting tool. Bullet containers should also be avoided (158).

This discussion of low temperature problems has focused on stocktype selection. The reader should be aware, however, of other practices that may moderate low temperatures and therefore affect seedling performance. Leaving overstory trees on the site, microsite planting, and leaving the litter layer relatively intact following harvest, may all contribute to improved thermal regimes.

9.4.3.5 Light. When considering the effects of light on tree response, it is essential to remember that light intensity, quality, and duration all influence growth (25, 42). Under natural conditions, however, light quality rarely has any significant effect on seedling performance (59 cited in 70), although numerous studies have shown light intensity (23, 52, 60) and duration (e.i. photoperiod, summarized in 124) to be important. The amount of literature dealing with these effects on tree physiology and growth is large. Several good, introductory references are recommended (86, 115, 116, 124, 151).

Species differ in their light requirements for minimum and optimum growth (70) which reflects differences in shade tolerance or light compensation and saturation points. Discussions of this can be found in several references (29, 42, 70, 116, 189, 204). These limits must be considered in preparing the reforestation prescription. For example, Ronco (170, 171) found that shade-tolerant Engelmann spruce (Picea engelmanii Parry) seedlings planted in full sun at high elevation suffered from solarization which caused high mortality. Water deficits during the first growing season following planting may have been an important factor, however. Once established, Engelmann spruce does quite well in full sun. On the other hand most shade-intolerant species, such as those of the genius Pinus, do better at higher light intensities when planted (provided there is adequate root growth and soil moisture). Species of intermediate shade-tolerance, such as Douglas-fir, may reach optimum growth in partial light that is 50 to 60 percent of full sunlight (23, 176). Situations in which light is the primary limiting factor are unusual. For this reason, few reports comparing the response of different stocktypes to varying degrees of light intensity have been published. An exception is work done by Iverson and Newton (101). They found that on sites dominated by coastal Oregon brush where moisture was not limiting, large bareroot 1-2 transplant Douglas-fir seedlings survived and produced more height growth than 2-0 bareroot, 1-0 plug, and bullet seedlings.

9.4.3.6 Nutrient deficiencies. The reforestation of sites with limiting nutrient deficiencies has been approached primarily at the species level. The rationale is that only those species adapted to specific nutrient problem sites should be used. For example, Jeffrey pine (Pinus jeffreyi Grev. & Balf.) is more productive than ponderosa pine on serpentine soils because of a greater tolerance of high levels of chromium, nickel, and magnesium and deficiencies of molybdenum and calcium (212). It has also been suggested that specific genotypes within a species respond differently to various concentrations of mineral nutrients (140).

Nutrient deficiencies affect seedling physiology in various ways depending upon the nutrient involved. The most obvious manifestation of a nutrient deficiency is reduced growth. Nutrient excesses on the other hand are unusual in the forest environment (116). For a general discussion of nutrients and physiology, the reader is referred to Kramer and Kozlowski (116).

In terms of stocktype selection, a considerable body of evidence has accumulated to show that mycorrhizae can improve nutrient absorption (18, 78, 118). For example, it was found that endomycorrhizae greatly improved nutrient absorption in red maple (Acer rubrum L) grown in anthracite waste (41). Guidelines for the use of site-specific fungal symbionts remain to be developed, however, for many types of sites where key nutrients such as nitrogen, phosphorus, and potassium are deficient in available forms.

9.4.3.7 Animal damage. Many different animal species severely damage or kill planted seedlings each year. These include elk, bear, deer, mountain beavers, hares, rabbits, birds, porcupines, squirrels, pocket gophers, wood rats, mice, and other animals common to the forest environment. Domestic livestock also occasionally damage regeneration (15). Mortality from animal damage is usually only severe at the local level and may vary in intensity from site-to-site as well as from year-to-year. Damage can be particularly severe even if mortality does not result, causing significant reductions in height growth (14). More frequently, however, animal damage is of secondary importance behind abiotic factors such as water and heat. There are always notable exceptions such as the high mortality in new plantations from pocket gophers in central Oregon (11).

There is ample evidence, either direct or implied, to suggest that large seedlings that are capable of producing rapid height and diameter growth are less likely to be destroyed by animals and are better able to recover than smaller seedlings (11, 15, 31, 47, 79, 101, 198). I have also observed that

1-0 container-grown Douglas-fir seedlings are much more susceptible to browse damage by deer than are 2-0 bareroot or plug-1 transplant stock. Genetypic differences can also affect selection by some animal pests in certain tree species (48, 166), although resistance to animal damage does not seem to be a major objective of tree improvement programs.

9.4.4 Morphology and stock quality

For many years seedling morphological characteristics such as height, diameter, and shoot-root ratio, were thought to determine seedling performance in the field. Although form and structure are still considered significant factors in seedling ability to tolerate environmental stress, high stock quality (physiological vigor) is now recognized as a precondition for good survival and growth. (A 1980 issue of the New Zealand Journal of Forestry Science, Vol. 10, No. 1, has been devoted to the topic of stock quality.) This is not meant to imply, however, that morphological characteristics are not important since it has already been shown in previous sections that specific morphologies can affect field performance. Ultimately, both physiological and morphological characteristics are important to the seedling's ability to withstand various types of stress. Discussions of these attributes are contained in several recent publications (29, 31, 167, 180).

9.5 CURRENT METHODS OF SELECTION

There are numerous systems in use today for the selection of species to be planted. Few systems, however, have been developed for stocktype selection and almost none are based on a systematic quantitative evaluation of planting alternatives over a variety of site conditions. Some guidelines are very general, providing a sequential list of decision steps that culminate in selection of the species or stocktype to be planted (30, 32, 164). Others discuss expected species or stocktype performance in certain types of environments (29, 106, 183, 186). Canadian researchers in British Columbia have taken a novel approach to the problem by the development of a system that utilizes nomograms incorporating economic as

well as seedling performance factors (208). One of the
unique features of this system is that variability in
seedling performance is considered.

9.6 A CONCEPTUAL SELECTION MODEL

A quantitative approach which carefully examines seedling
performance in relation to the operational environment is
rare in practice. All too often, particularly at the stocktype level, reliance is placed on blanket prescriptions with
little regard for differential response to environmental
change. Lacking is a methodology to systematically evaluate
species and stocktype alternatives for a range of operational
environments in terms of seedling performance. The development of such a system, to provide site-specific species and
stocktype guidelines, would obviously require a strong commitment from management.

A conceptual model of one such system would require the
completion of several developmental stages (Figure 2). The
necessary framework upon which the evaluation is based is an
environmental classification of the lands under study in terms
of those operational factors considered most limiting.

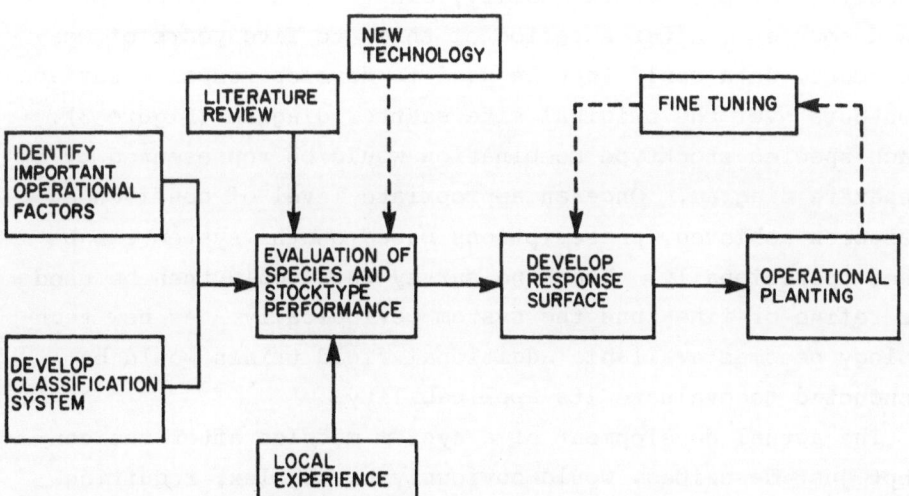

FIGURE 2. Conceptual model of a system that could be used to
identify species and stocktypes suitable for specific operational environments.

Obviously, it is not practical to expect to be able to directly measure operational factors on a large number of sites
over a period of several years. Approximations can be made
indirectly, however, from such things as precipitation maps,
soil water holding capacity estimates, number of frost-free
days, calculation of potential direct beam solar radiation,
elevation, and others. Individual sites can then be located
on a scatter diagram of from one to three dimensions with
each axis representing one operational factor. This in effect
is a form of gradient analysis. Three dimensional diagrams,
however, should probably be avoided if possible because of
their added complexity. Once this has been accomplished field
trials may be installed with spatial and temporal replication
arranged so that as much site variation (i.e., operational
factor variation) can be sampled as is realistic. Treatments
selected for field trial inclusion should be identified from
a review of pertinent literature and local experience. Since
the major objective is to compare treatment responses under
various environmental conditions, it is important that outside
sources of variation be minimized. These would include such
things as inadequate site preparation, planting during unfavorable weather, poor seedling handling and planting, using
seedlings of poor stock quality, and the use of inappropriate
seed sources. After a period of three to five years of measurement, data could then be used to develop response surface
contours over the original site scatter diagram (Figure 3).
Each species-stocktype combination would be represented by a
separate diagram. Once an appropriate level of confidence
has been achieved, prescriptions based on the system can be
used operationally. Stocking survey data could then be used
to refine or fine-tune the system periodically. As new technology becomes available additional field trials would be
conducted to evaluate its applicability.

The actual development of a system modeled after the concept just described, would obviously be complex, requiring
descriptive site data on all units recognized as individual
entities (e.g., a timber stand), and the probable use of
sophisticated statistical procedures. A system based on this

conceptual model is now being tested in southwest Oregon (84).
If site-specific prescription guidelines are to take advantage
of specific species and stocktype characteristics, then a
careful evaluation of seedling response to changing environmental conditions is necessary.

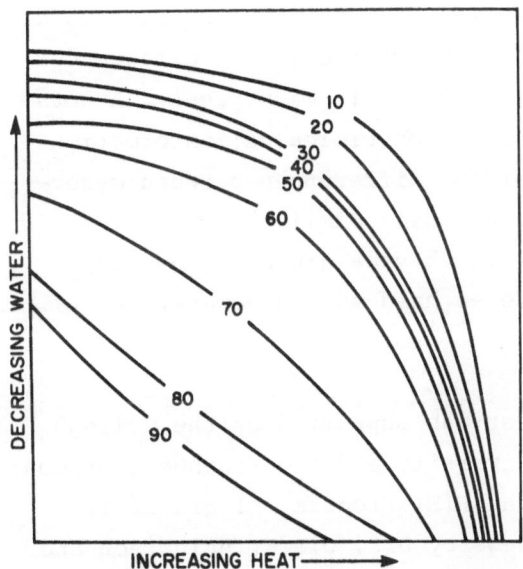

FIGURE 3. Hypothetical depiction of seedling survival (%)
for a single species and stocktype under various moisture and
heat conditions found within a sample area.

9.7 CONCLUSIONS

A prerequisite for successful reforestation is the identification of those factors in the operational environment
that exert the strongest influences on seedling survival and
growth. This, and an understanding of response differences
between species and stocktype planting options, will improve
the forester's ability to match seedling genetic, morphological, and physiological characteristics to site conditions.
Although there are a number of methodologies in use today for
species and stocktype selection, none are designed to systematically evaluate response differences along environmental
gradients. Such a system would be particularly useful in
areas characterized by extremes of environmental stress.

In the last 25 years tremendous progress has been made in
our understanding of ecophysiological differences between
species in response to environmental stress. Knowledge of
the changes in growth and physiology of individual stocktypes
subjected to various types of stress has also advanced during
the same period. Numerous trials of different stocktypes
have been conducted on a wide variety of sites, but our understanding of the physiological bases for observed differences
in survival and growth remain obscure in many cases. A notable exception is the vast body of physiological information
that has been developed regarding differences between mycorrhizal and nonmycorrhizal seedlings. Similarly, detailed
comparative studies of other stocktypes are badly needed,
particularly on sites considered hard-to-regenerate.

ACKNOWLEDGEMENTS

Preparation of this chapter was supported by the College
of Forestry, Oregon State University and the Southwest Oregon
Forestry Intensified Research (FIR) Program. I gratefully
acknowledge the helpful reviews by Drs. Ole T. Helgerson and
Steven D. Tesch, Department of Forest Science, and Mr. John
W. Mann, Department of Forest Engineering, Oregon State
University. Special thanks to Mrs. Elaine Morse for help
with the numerous revisions.

REFERENCES
1. Aldhous, J. R. 1972. Nursery practice. United Kingdom
    Comm. Bull. 43, HMSO, London.
2. Alm, A. A. 1983. Black and white spruce plantings in
    Minnesota: Container vs bareroot stock and fall vs spring
    planting. For. Chron. 59:189-191.
3. Amidon, T. E., J. P. Barnett, H. P. Gallagher and J. M.
    McGilvray. 1982. A field test of containerized seedlings
    under drought conditions. pp. 139-144 In R. W. Guldin,
    and J. P. Barnett (Eds.). Proc., Southern containerized
    forest tree seedling conference. USDA For. Ser. Gen.
    Tech. Rep. SO-37, New Orleans. 156 p.
4. Anderson, M. L. 1950. The selection of tree species:
    An ecological basis of site classification for conditions
    found in Great Britian and Ireland. Oliver and Boyd,
    Edinburg.
5. Anderson, J. M., and R. C. Williamson. 1974. Use of
    germinated seed to establish Coulter pine plantations in
    southern California. J. For. 72: 90-91.

6. Anonymous. 1981. 1981 directory of forest tree nurser-
    ies in the United States. Amer. Assoc. of Nurserymen
    and USDA For. Ser. 40 p.
7. Armson, K. A., and V. Sadreika. 1974. Forest tree nur-
    sery soil management and related practices. Min. of
    Nat. Res., Div. of For., Forest Management Branch,
    Ontario. 177 p.
8. Arnott, J. T. 1975. Field performance of container-
    grown and bareroot trees in coastal British Columbia.
    Can. J. For. Res. 5:186-194.
9. Baer, H. W., and J. D. Otta. 1981. Outplanting survival
    and growth of ponderosa pine seedlings inoculated with
    Pisolithus tinctorius in South Dakota. For. Sci.
    27:277-280.
10. Baker, F. S. 1934. Theory and practice of silviculture.
    McGraw-Hill Book Company, Inc., New York. 502 p.
11. Barnes, V. G., Jr. 1978. Survival and growth of ponder-
    osa pine seedlings injured by pocket gophers. Tree
    Planters' Notes 29(2)20-23.
12. Barrett, J. W. (Ed.). 1962. Regional silviculture of
    the United States. The Ronald Press Company, New York.
    610 p.
13. Bilan, M. V., and S. W. Jan. 1968. Needle moisture
    content as indicator of cessation of root elongation in
    loblolly pine seedlings. Bull. Ecol. Soc. Amer. 49:109.
14. Black, H. C., E. J. Dimock II, W. E. Dodge, and W. H.
    Lawrence. 1969. Survey of animal damage on forest plan-
    tations in Oregon and Washington. pp. 388-408 In Proc.,
    Thirty-fourth North Amer. wildlife and nat. res. conf.
    Wildlife Mgt. Inst., Washington, D.C.
15. Black, H. C., E. J. Dimock II, J. Evans, and J. A.
    Rochelle. 1979. Animal damage to coniferous plantations
    in Oregon and Washington. Part 1. A survey, 1963-1975.
    For. Res. Lab. Res. Bull. 25. Oreg. State Univ.,
    Corvallis. 45 p.
16. Blackman, F. F., and G. L. C. Matthaei. 1905. Experi-
    mental researches in vegetable assimilation and respira-
    tion. IV. A quantitative study of carbon dioxide assim-
    ilation and leaf temperature in natural illumination.
    Proc. Royal Soc. London, Ser. B. 76:402-460.
17. Bledsoe, C. S., K. Tennyson, and W. Lopushinsky. 1982.
    Survival and growth of outplanted Douglas-fir seedlings
    inoculated with mycorrhizal fungi. Can. J. For. Res.
    12:720-723.
18. Bowen, G. D. 1973. Mineral nutrition of ectomycorrhizae.
    pp. 151-205 In C. G. Marks and T. T. Kozlowski (Eds.).
    Ectomycorrhizae - their ecology and physiology. Academic
    Press, New York.
19. Boyce, J. S. 1961. Forest pathology. McGraw-Hill Book
    Company, Inc., New York. 572 p.
20. Boyer, J. S. 1976. Water deficits and photosynthesis.
    pp. 153-190 In T. T. Kozlowski (Ed.). Water deficits
    and plant growth. Vol. IV. Academic Press, New York.
21. Briscoe, C. B. 1969. Establishment and early care of
    sycamore plantations. USDA For. Serv. Res. Paper SO-50.
    Southern For. Exp. Sta., New Orleans. 18 p.

22. Brix, H. 1962. The effect of water stress on the rates of photosynthesis and respiration in tomato plants and loblolly pine seedlings. Physiol. Plant. 15:10-20.

23. Brix, H. 1967. An analysis of dry matter production of Douglas-fir seedlings in relation to temperature and light intensity. Can J. Bot. 45:2063-2072.

24. Brix, H. 1979. Effects of plant water stress on photosynthesis and survival of four conifers. Can. J. For. Res. 9:160-165.

25. Brown, C. L. 1971. Growth and form. pp. 125-167 In M. H. Zimmermann and C. L. Brown (Eds.). Trees: Structure and function. Springer-Verlag, New York.

26. Burdett, A. N. 1979. New methods for measuring root growth capacity: Their value in assessing lodgepole pine stock quality. Can. J. For. Res. 9:63-67.

27. Champion, H., and N. V. Brasnett. 1959. Choice of tree species. FAO Forestry Development Paper No. 13. Food and Agriculture Organization of the United Nations, Rome. 307 p.

28. Cleary, B.D. 1971. The effect of plant moisture stress on the physiology and establishment of planted Douglas-fir and ponderosa pine seedlings. Ph.D. Thesis, Oreg. State Univ., Corvallis. 85 p.

29. Cleary, B. D., and R. D. Greaves. 1976. Determining planting stock needs. pp. 60-81 In D. M. Baumgartner and R. J. Boyd (Eds.). Tree planting in the Inland Northwest. Wash. State Univ. Coop. Ext. Serv., Pullman.

30. Cleary, B. D., and R. D. Greaves. 1978. The reforestation plan. pp. 163-186 In B. D. Cleary, R. D. Greaves, and R. K. Hermann (Eds.). Regenerating Oregon's forests Oreg. State Univ. Ext. Ser., Corvallis.

31. Cleary, B. D., R. D. Greaves, and P. W. Owston. 1978. Seedlings. pp. 63-97 In B. D. Cleary, R. D. Greaves, and R. K. Hermann (Eds.). Regenerating Oregon's forests. Oreg. State Univ. Ext. Serv., Corvallis.

32. Cleary, B. D., and B. R. Kelpsas. 1981. Five steps to successful regeneration planning. For Res. Lab. Spec. Publ. 1. Oreg. State Univ., Corvallis. 31 p.

33. Cleary, B. D., and R. H. Waring. 1969. Temperature: Collection of data and its analysis for the interpretation of plant growth and distribution. Can. J. Bot. 47:167-173.

34. Cochran, P. H. 1972. Tolerance of lodgepole and ponderosa pine seeds and seedlings to high water tables. Northwest Sci. 46:322-331.

35. Cochran, P. H. 1975. Soil temperatures and natural regeneration in south-central Oregon. pp. 37-52 In B. Bernier and C. H. Winget (Eds.). Forest soils and forest land management. Les Presses de L'Universite' Laval, Quebec.

36. Cochran, P. H., and C. M. Berntsen. 1973. Tolerance of lodgepole and ponderosa pine seedlings to low night temperatures. For. Sci. 19:272-280.

37. Cordell, C. E., and D. H. Marx. 1981. Ectomycorrhizae: Present status and practical application in forest tree nurseries and field plantings. pp. 34-39 In Proc.,

Intermountain nurseryman's assoc. and western forest
nursery assoc. meeting. USDA For. Serv. Gen. Tech. Rep.
INT-109.

38. Coutts, M. P., and W. Armstrong. 1976. Role of oxygen
transport in the tolerance of trees to waterlogging.
pp. 361-385 In M. G. R. Cannell and F. T. Last (Eds.).
Tree physiology and yield improvement. Academic Press,
New York.

39. Crafts, A. S. 1968. Water deficits and physiological
processes. pp. 85-133 In T. T. Kozlowski (Ed.). Water
deficits and plant growth. Vol. II. Plant water consumption and response. Academic Press, New York.

40. Crawford, R. M. M. 1976. Tolerance of anoxia and the
regulation of glycolysis in tree roots. pp. 387-401 In
M. G. R. Cannell and F. T. Last (Eds.). Tree physiology
and yield improvement. Academic Press, New York.

41. Daft, M. J., and E. Hacskaylo. 1977. Growth of endomycorrhizal and nonmycorrhizal red maple seedlings in
sand and anthracite spoil. For. Sci. 23:207-216.

42. Daniel, T. W., J. A. Helms, and F. S. Baker. 1979.
Principles of silviculture. McGraw-Hill Book Company,
Inc., New York. 500 p.

43. Daubenmire, R. 1976. The use of vegetation in assessing
the productivity of forest lands. Bot. Rev. 42:115-143.

44. Daubenmire, R. 1978. Plant geography. Academic Press,
New York. 339 p.

45. Davidson, W. H., and E. A. Sowa. 1974. Early attempts
to vegetate coal-mine spoils with container-grown seedlings. pp. 372-376 In R. W. Tinus, W. I. Stein, and
W. E. Blamer (Eds.). Proc., North Amer. containerized
forest tree seedling symp. Great Plains Agric. Counc.
Publ. 68.

46. DeVelice, R. L., and B. A. Buchanan. 1978. Germinant
reforestation: A promising new technique. Tree Planters'
Notes 29:3-6.

47. Dimock, E. J., II. 1970. Ten-year height growth of
Douglas-fir damaged by hare and deer. J. For. 68:285-288.

48. Dimock, E. J., II, R. R. Silen, and V. E. Allen. 1976.
Genetic resistance in Douglas-fir to damage by snowshoe
hare and black-tailed deer. For. Sci. 22:106-121.

49. Dixon, R. K., D. E. Garrett, and G. S. Cox. 1981. Con-
tainer- and nursery-grown black oak seedlings inoculated
with Pisolithus tinctorius: Growth and ectomycorrhizal
development following outplanting on an Ozark clearcut.
Can. J. For. Res. 11:492-496.

50. Dixon, R. K., S. G. Pallardy, H. E. Garrett, and G. S.
Cox. 1983. Comparative water relations of container-
grown and bareroot ectomycorrhizal and nonmycorrhizal
Quercus velutina seedlings. Can. J. Bot. 61:1559-1565.

51. Dixon, R. K., G. M. Wright, G. T. Behrns, R. O. Teskey,
and T. M. Hinckley. 1980. Water deficits and root
growth of ectomycorrhizal white oak seedlings. Can. J.
For. Res. 10:545-548.

52. Drew, A. P., and W. K. Ferrell. 1977. Morphological
acclimation to light intensity in Douglas-fir seedlings.
Can. J. Bot. 55:2033-2042.

53. Duryea, M. L., and T. D. Landis. (Eds.). 1984. Forest nursery manual: Production of bareroot seedlings. Martinus Nijhoff/Dr W. Junk Publishers, The Hague, Netherlands, for Oreg. State Univ., For. Res. Lab. 340 p.

54. Duryea, M. L, and D. P. Lavender. 1982. Water relations, growth, and survival of root-wrenched Douglas-fir seedlings. Can. J. For. Res. 12:545-555.

55. Edgren. J. W. 1968. Douglas-fir nursery stock survival by age class. pp. 52-56 In Proc., West. For. and Conserv. Assoc. and West. For. Nursery Counc.

56. Edgren, J. W. 1970. Growth of frost-damaged Douglas-fir seedlings. USDA For. Serv. Res. Note PNW-121. Pac. Northwest For. and Range Exp. Sta., Portland. 8 p.

57. Edgren, J. W. 1977. Field survival and growth of Douglas-fir by age and size of nursery stock. USDA For. Serv. Res. Paper PNW-217. Pac. Northwest For. and Range Exp. Sta., Portland. 6 p.

58. Edgren, J. W. 1981. Field performance of undercut coastal and Rocky Mountain Douglas-fir 2+0 seedlings. Tree Planters' Notes 32:33-36.

59. Edgren, J. M., and W. I. Stein. 1974. Artificial regeneration. pp. M1-M32. In O. P. Cramer (Ed.). Environmental effects of forest residues management in the Pacific Northwest, a state-of-knowledge compendium. USDA For. Serv. Gen. Tech. Rep. PNW-24. Pac. Northwest For. and Range Exp. Sta., Portland.

60. Emmingham, W. H., and R. H. Waring. 1973. Conifer growth under different light environments in the Siskiyou Mountains of southwestern Oregon. Northwest Sci. 47:88-99.

61. Foiles, M. W., and J. D. Curtis. 1973. Regeneration of ponderosa pine in the northern Rocky Mountain-Intermountain Region. USDA For. Serv. Res. Paper INT-145. Intermountain For. and Range Exp. Sta., Ogden. 44 p.

62. Fowells, H. A. 1965. Silvics of forest trees of the United States. USDA For. Serv. Agric. Handb. 271, Washington, D.C. 762 p.

63. Furlan, V., J. A. Fortin, and C. Plenchette. 1983. Effects of different vesicular-arbuscular mycorrhizal fungi on growth of Fraxinus americana. Can. J. For. Res. 13:589-593.

64. Gill, C. J. 1970. The flooding tolerance of woody species - a review. Forest. Abstr. 31:671-688.

65. Gilmore, A. R. 1980. Extending the range of loblolly pine in the Mississippi River Valley: Factors relating to growth and longevity. pp. 8-14 In Proc., 1980 southern nursery conference.

66. Gilmore, A. R., and G. E. Metcalf. 1961. Site quality curves for plantation-grown loblolly pine in southern Illinois. Forestry Note No. 97, Univ. of Ill. Agr. Exp. Sta., Urbana. 2 p.

67. Ginter-Whitehouse, D. L., T. M. Hinckley, and S. G. Pallardy. 1983. Spatial and temporal aspects of water relations of three tree species with different vascular anatomy. For. Sci. 29:317-329.

68. Glerum, C. 1976. Frost hardiness of forest trees. pp. 403-420 In M. G. R. Cannell and F. T. Last (Eds.). Tree physiology and yield improvement. Academic Press, New York.

69. Graham, S. A. 1952. Forest entomology. Third Edition. McGraw-Hill Book Company, Inc., New York.

70. Greaves, R. D., R. K. Hermann, and B. D. Cleary. 1978. Ecological principles. pp. 7-26 In B. D. Cleary, R. D. Greaves, and R. K. Hermann (Eds.). Regenerating Oregon's forests. Oreg. State Univ. Ext. Serv., Corvallis.

71. Guldin, R. W. 1982. What does it cost to grow seedlings in containers? Tree Planters' Notes 33(1):34-37.

72. Guldin, R. W., and J. P. Barnett (Eds.). 1981. Proc., Southern containerized forest tree seedlings conf. USDA For. Serv. Gen. Tech. Rep. SO-37. Southern For. Exp. Sta., New Orleans. 156 p.

73. Hahn, P. F. 1981. A historical overview of the use of containerized seedlings for operational reforestation. - How did we get where we are today? pp. 7-12 In R. W. Guldin, and J. P. Barnett (Eds.). Proc., Southern containerized forest tree seedling conf. USDA For. Serv. Gen. Tech. Report SO-37. Southern For. Exp. Sta., New Orleans.

74. Hahn, P. F., and A. J. Smith. 1983. Douglas-fir planting stock performance comparison after the third growing season. Tree Planters' Notes 34(1)33-39.

75. Hansen, E. A., and H. M. Phipps. 1983. Effect of soil moisture tension and preplant treatments on early growth of hybrid Populus hardwood cuttings. Can. J. For. Res. 13:458-464.

76. Hansen. E. A., and D. N. Tolsted. 1981. Effect of cutting diameter and stem or branch position on establishment of a difficult-to-root clone of a Populus alba hybrid. Can. J. For. Res. 11:723-727.

77. Hare, R. C., and S. B. Land, Jr. 1982. Effect of cold storage and chemical treatment on rooting of hardwood sycamore cuttings. Can. J. For. Res. 12:417-419.

78. Harley, J. L. 1969. The biology of mycorrhizae. Leonard Hill, London. 334 p.

79. Hartwell, H. D. 1969. Control of damage by snowshoe hares on forest plantations. pp. 80-81 In H. C. Black (Ed.). Wildlife and reforestation in the Pacific Northwest. For. Res. Lab., Oreg. State Univ., Corvallis.

80. Heidmann, L. J. 1976. Frost heaving of tree seedlings: A literature review of causes and possible controls. USDA For. Serv. Gen. Tech. Rep. RM-21, Rocky Mtn. For. and Range Exp. Sta., Ft. Collins. 10 p.

81. Heidmann, L. J., and D. B. Thorud. 1976. Controlling frost heaving of ponderosa pine seedlings in Arizona. USDA For. Serv. Res. Paper RM-172., Rocky Mtn. For. and Range Exp. Sta., Ft. Collins. 12 p.

82. Heilman, P. E., and G. Ekuan. 1979. Effect of planting stock length and spacing on growth of black cottonwood. For. Sci. 25:439-443.

83. Helgerson, O. T. 1981. Nitrogen fixation by scotch broom (Cytisus scoparius L.) and red alder (Alnus ruba

Bong.) planted under precommercially thinned Douglas-fir [Pseudotsuga menziesii (Mirb.) Franco]. Ph.D. Thesis, Oreg. State Univ., Corvallis. 81 p.

84. Helgerson, O. T. 1983. Personal communication. Dept. of For. Sci., Oreg. State Univ., Corvallis.

85. Helgerson, O. T., 1983. Unpublished data on file at the Dept. of For. Sci. Data Bank, Oreg. State Univ., Corvallis.

86. Helms, J. A. 1976. Factors influencing net photosynthesis in trees: An ecological viewpoint. pp. 55-78 In M. G. R. Cannell and F. T. Last (Eds.). Tree physiology and yield improvement. Academic Press, New York.

87. Heninger. R. L., and D. P. White. 1974. Tree seedling growth at different soil temperatures. For. Sci. 20:363-367.

88. Hermann, R. K. 1964. Importance of top-root ratios for survival of Douglas-fir seedlings. Tree Planters' Notes 64:7-11.

89. Hermann, R. K. 1965. Survival of planted ponderosa pine in southern Oregon. For. Res. Lab. Res. Paper 2, Oreg. State Univ., Corvallis. 32 p.

90. Hermann, R. K., and D. P. Lavender. 1968. Early growth of Douglas-fir from various altitudes and aspects in southern Oregon. Silvae Genet. 17:143-151.

91. Hermann, R. K., D. P. Lavender, and J. B. Zaerr. 1972. Lifting and storing western conifer seedlings. For. Res. Lab. Res. Paper 17, Oreg. State Univ., Corvallis. 8 p.

92. Hobbs, S. D. 1982. Performance of artifically shaded container-grown Douglas-fir seedlings on skeletal soils. For. Res. Lab. Res. Note 71, Oreg. State Univ., Corvallis. 6 p.

93. Hobbs, S. D. 1982. Stocktype selection and planting techniques for Douglas-fir on skeletal soils in southwest Oregon. pp. 92-96 In S. D. Hobbs and O. T. Helgerson (Eds.). Proc., Reforestation of skeletal soils workshop. For. Res. Lab., Oreg. State Univ., Corvallis.

94. Hobbs, S. D., D. P. Lavender, and K. A. Wearstler, Jr. 1982. Performance of container-grown Douglas-fir on droughty sites in southwest Oregon. pp. 373-377 In J. B. Scarratt, C. Glerum, and C. A. Plexman (Eds.). Proc., Canadian containerized tree seedling symp. Can. For. Serv., Saulte Ste. Marie, Ontario.

95. Hobbs, S. D., and K. A Wearstler, Jr. 1983. Performance of three Douglas-fir stocktypes on a skeletal soil. Tree Planters' Notes 34(3):11-14.

96. Hobbs, S. D., and S. G. Stafford. Unpublished data on file at the Dept. of For. Sci. Data Bank, Oreg. State Univ., Corvallis.

97. Hocker, H. W., Jr. 1979. Introduction to forest biology. John Wiley and Sons, Inc., New York. 467 p.

98. Hook, D. D., and C. L. Brown. 1973. Root adaptations and relative flood tolerance of five hardwood species. For. Sci. 19:225-229.

99. Hosner, J. F., and S. G. Boyce. 1962. Relative

tolerance to water saturated soil of various bottomland hardwoods. For. Sci. 8:180-186.

100. Hunt, R., and B. Zobel. 1978. Frost-hardy Eucalyptus grow well in the southeast. S. J. Appl. For. 2(1)6-10.

101. Iverson, R. D., and M. Newton. 1980. Large Douglas-fir seedlings perform best on Oregon coastal sites. International Paper Company Tech. Note 55. Western Forest Research Center. Lebanon, Oregon. 9 p.

102. Jacobs, M. R. 1979. Eucalypts for planting. FAO Forestry Series No. 11, Food and Agriculture Organization of the United Nations, Rome. 677 p.

103. Jenkinson, J. L., and J. A. Nelson. 1982. 1-0 Douglas-fir: A bareroot planting option. pp. 63-76 In Proc., Western forest nursery council.

104. Johnson, P. S. 1981. Early results of planting English oak in an Ozark clearcut. USDA For. Serv. Res. Paper NC-204. North Central For. Exp. Sta., St. Paul. 6 p.

105. Kellison, R. C., R. D. Heeren, and S. Jones. 1978. Species selection as related to soils in the Atlantic Coastal Plain. pp. 42-48 In Proc., 6th southern soils forest workshop. USDA For. Serv., Southeast Area., State and Priv. For.

106. Kellison, R. C., and J. B. Jett, Jr. 1978. Species selection for plantation establishment in the Atlantic Coastal Plain and Sandhills Provinces. pp. 196-202 In W. E. Balmer (Ed.). Proc., Soil moisture...site productivity symp. USDA For. Serv., Southeast Area, State and Priv. For.

107. Kinghorn, J. M. 1974. Principles and concepts in container planting. pp. 8-18 In R. W. Tinus, W. I. Stein, and W. E. Balmer (Eds.). Proc., North Amer. containerized forest tree seedling symp. Great Plains Agric. Counc. Publ. 68.

108. Koon, K. B., and T. O'Dell. 1977. Effects of wrenching on drought avoidance of Douglas-fir seedlings. Tree Planters' Notes 28:15-16.

109. Kormanik, P. P., W. C. Bryan, and R. C. Schultz. 1977. Influence of endomycorrhizae on growth of sweetgum seedlings from eight mother trees. For. Sci. 23:500-505.

110. Kozlowski, T. T. 1968. Introduction. pp. 1-11 In T. Kozlowski (Ed.). Water deficits and plant growth. Vol. I. Academic Press, New York.

111. Kozlowski, T. T. 1971. Growth and development of trees. Vol. I. Seed germination, ontogeny, and shoot growth. Academic Press, New York. 443 p.

112. Kozlowski, T. T. 1976. Water supply and leaf shedding. pp. 191-231 In T. T. Kozlowski (Ed.). Water deficits and plant growth. Vol. IV. Academic Press, New York.

113. Kramer, P. J. 1964. The role of water in wood formation. pp. 519-532 In M. H. Zimmermann (Ed.). The formation of wood in forest trees. Academic Press, New York.

114. Kramer, P. J. 1969. Plant and soil water relationships: A modern synthesis. McGraw-Hill Book Company, Inc., New York. 482 p.

115. Kramer, P. J., and T. T. Kozlowski. 1960. Physiology

of trees. McGraw-Hill Book Company, Inc., New York. 642 p.

116. Kramer, P. J., and T. T. Kozlowski. 1979. Physiology of woody plants. Academic Press, New York. 811 p.

117. Krinard, R. M. 1983. Continued investigations in first-year survival of long cottonwood cuttings. Tree Planters' Notes 34:34-37.

118. Lamont, B. 1982. Mechanisms for enhancing nutrient uptake in plants, with particular reference to Mediterranean South Africa and western Australia. The Bot. Rev. 48:597-689.

119. Land, S. B., Jr. 1983. Performance and G-E interactions of sycamore established from cuttings and seedlings. pp. 431-440 In E. P. Jones, Jr. (Ed.). Proc., Second biennial southern silviculture res. conf. USDA For. Serv. Gen. Tech. Report SE-24. Southeastern For. Exp. Sta., Ashville.

120. Langlois, C. G., and J. A. Fortin. 1982. Mycorrhizal development on containerized tree seedlings. pp. 183-202 In J. B. Scarratt, C. Glerum, and C. A. Plexman (Eds.). Proc., Canadian containerized tree seedling symp. Can. For. Serv., Great Lakes Forest Research Centre, Sault Ste. Marie.

121. Larson, M. M. 1961. Seed size, germination dates, and survival relationships of ponderosa pine in the Southwest. USDA For. Serv. Res. Note 66. Rocky Mtn. For. and Range Exp. Sta., Ft. Collins. 4 p.

122. Larsson, S., R. Oren, R. H. Waring, and J. W. Barrett. 1983. Attacks of mountain pine beetle as related to tree vigor of ponderosa pine. For. Sci. 29:395-402.

123. Lavender, D. P. 1964. Date of lifting for survival of Douglas-fir seedlings. For. Res. Lab. Res. Note 49, Oreg. State Univ., Corvallis. 20 p.

124. Lavender, D. P. 1981. Environment and shoot growth of woody plants. For. Res. Lab. Res. Paper 45, Oreg. State Univ., Corvallis. 47 p.

125. Lavender, D. P., and B. D. Cleary. 1974. Coniferous seedling production techniques to improve seedling establishment. pp. 177-180 In R. W. Tinus, W. I. Stein, and W. E. Balmer (Eds.). Proc., North Amer. containerized forest tree seedling symp., Great Plains Agric. Counc. Publ. 68.

126. Leshem, B. 1970. Resting roots of Pinus halepensis: Structure, function and reaction to water stress. Bot. Gaz. 131:99-104.

127. Levitt, J. 1980. Response of plants to environmental stresses. Vol. I. Chilling, freezing and high temperature stresses. Academic Press, New York. 497 p.

128. Liebig, J. von. 1855. Die Grundsatze der Agriculturchemie, mit Reecksicht auf die in England angestellten Untersuchungen. 152 s. F. Vieveg und Sohn, Braunschweig.

129. Lopushinsky, W., and T. Beebe. 1976. Relationship of shoot-root ratio to survival and growth of outplanted Douglas-fir and ponderosa pine seedlings. USDA For. Serv. Res. Note PNW-274. Pac. Northwest For. and Range Exp. Sta., Portland. 7 p.

130. Lopushinsky, W., and G. O. Klock. 1974. Transpiration of conifer seedlings in relation to soil water potential. For. Sci. 20:181-186.

131. Maisenhelder, L. C., and J. S. McKnight. 1968. Cottonwood seedlings best for sites subject to flooding. Tree Planters' Notes 19(3):15-16.

132. Marks, C. G., and T. T. Kozlowski. (Eds.). 1973. Ectomycorrhizae - their ecology and physiology. Academic Press, New York. 444 p.

133. Marx, D. H. 1972. The importance of mycorrhizae in forest nurseries. pp. 188-192 In Proc., Southeastern area forest tree nurserymen's conference. Miss. For. Comm., Greenville.

134. Marx, D. H., and J. P. Barnett. 1974. Mycorrhizae and containerized forest tree seedlings. pp. 85-92 In R. W. Tinus, W. I. Stein, and W. E. Balmer (Eds.). Proc., North Amer. containerized forest tree seedling symp. Great Plains Agric. Counc. Publ. 68.

135. Marx, D. H., and W. C. Bryan. 1971. Influence of ectomycorrhizae on survival and growth of aseptic seedlings of loblolly pine at high temperatures. For. Sci. 17:37-41.

136. Marx, D. H., W. C. Bryan, and C. E. Cordell. 1977. Survival and growth of pine seedlings with Pisolithus ectomycorrhizae after two years on reforestation sites in North Carolina and Florida. For. Sci. 23:363-373.

137. Marx, D. H., W. C. Bryan, and C. B. Davey. 1970. Influence of temperature on aseptic synthesis of ectomycorrhizae by Thelephera terrestris and Pisolithus tinctorius on loblolly pine. For. Sci. 16:424-431.

138. Marx, D. H., J. G. Mexal, and W. G. Morris. 1979. Inoculation of nursery seedbeds with Pisolithus tintorius spores mixed with hydromulch increases ectomycorrhizae and growth of loblolly pines. S. J. Appl. For. 3:175-178.

139. Mason, H. L., and J. H. Langenheim. 1957. Language analysis and the concept environment. Ecol. 38:325-340.

140. Mason, P. A., and J. Pelham. 1976. Genetic factors affecting the response of trees to mineral nutrients. pp. 437-448 In M. G. R. Cannell and F. T. Last (Eds.). Tree physiology and yield improvement. Academic Press, New York.

141. McDermott, R. E. 1954. Effects of saturated soil on seedling growth of some bottomland hardwood species. Ecol. 35:36-41.

142. McDonald, P. M., and R. D. Cosens. 1980. Survival and height growth of bareroot and container-grown seedlings in northern California. p. 75 In Exe. Sum. Proc., 1980 West. For. Conf., West. For. and Conserv. Assoc., Portland.

143. McKnight, J. S. 1970. Planting cottonwood cuttings for timber production in the south. USDA For. Serv. Res. Paper SO-60. Southern For. Exp. Sta., New Orleans. 17 p.

144. Mikola, P. 1973. Application of mycorrhizal symbiosis in forestry practice. pp. 383-411 In C. G. Marks and

220

T. T. Kozlowski (Eds.). Ectomycorrhizae—their ecology and physiology. Academic Press, New York.

145. Miller, N. F., L. E. Hinsley, and F. A. Blazich. 1982. Rooting of Fraser fir cuttings: Effects of postseverance chilling and photoperiod during rooting. Can. J. For. Res. 12:607-611.

146. Minore, D. 1968. Effects of artificial flooding on seedling survival and growth of six northwestern tree species. USDA For.Serv. Res. Note PNW-92. Pac. Northwest For. and Range Exp. Sta., Portland. 12 p.

147. Minore, D. 1970. Seedling growth of eight northwestern tree species over three water tables. USDA For. Serv. Res. Note PNW-115. Pac. Northwest For. and Range Exp. Sta., Portland. 8 p.

148. Minore, D. 1979. Comparative autecological character- istics of northwestern tree species...a literature review. USDA For. Serv. Gen. Tech. Rep. PNW-87. Pac. Northwest For. and Range Exp. Sta., Portland. 72 p.

149. Molina, R. 1977. Ectomycorrhizal fungi and forestry practice. pp. 147-161 In T. Walters (Ed.). Mushrooms and man, an interdisciplinary approach to mycology. Linn-Benton Community College, Albany, Oregon.

150. Molina, R. 1982. Mycorrhizal inoculation and its potential impact on seedling survival and growth in southwest Oregon. pp. 86-91 In S. D. Hobbs and O. T. Helgerson (Eds.). Proc., Reforestation of skeletal soils. For. Res. Lab., Oreg. State Univ., Corvallis.

151. Morison, J. I. L., and P. G. Jarvis. 1983. Direct and indirect effects of light on stomata. I. In Scots pine and Sitka spruce. Plant, Cell, and Environ. 6:95-101.

152. Mullin, R. E. 1980. Comparison of seedling and trans- plant performance following 15 years growth. For. Chron. 56:231-232.

153. Murry, M. D., and C. A. Harrington. 1983. Growth and yield of a 24-year-old black cottonwood plantation in western Washington. Tree Planters' Notes 34:3-5.

154. Newton, M. 1978. A test of western hemlock wildlings in brushfield reclamation. For. Res. Lab. Res. Paper 39, Oreg. State Univ., Corvallis. 22 p.

155. Nicholson, L. A. 1974. Seedling size description. pp. 74-76 In Proc., West. for. nursery counc. meeting.

156. Orlov, A. Y. 1980. Cyclic development of roots of conifers and their relation to environmental factors. pp. 43-61 In D. N. Sen (Ed.). Environment and root behavior. Geobios International, Jodhpur, India.

157. Outcalt, K. W. 1982. Selecting pine species for flatwoods sites. Tree Planters' Notes 33:18-19.

158. Owston. P. W. 1983. Personal communication. USDA For. Serv., Pac. Northwest For. and Range Exp. Sta., Corvallis, Oregon.

159. Owston, P. W., and K. W. Seidel. 1978. Container and root treatments affect growth and root form of planted ponderosa pine. Can. J. For. Res. 8:232-236.

160. Parke, J. L., R. G. Linderman, and C. H. Black. 1983. The role of ectomycorrhizae in drought tolerance of Douglas-fir seedlings. New Phytol. 95:83-95.

161. Peace, T. R. 1961. The dangerous concept of the natural forest. Quarterly J. For. 55:12-23.

162. Parmeter, J. R., Jr. 1978. Forest stand dynamics and ecological factors in relation to dwarf mistletoe spread, impact, and control. pp. 16-30 In R. F. Scharpf and J. R. Parmeter, Jr. (Eds.). Proc., Symp. on dwarf mistletoe control through forest management. USDA For. Serv. Gen. Tech. Report PSW-31. Pac. Southwest For. and Range Exp. Sta., Berkeley.

163. Petersen, G. J. 1982. The effects of artificial shade on seedling survival on western Cascade harsh sites. Tree Planters' Notes 33(1):20-23.

164. Pfister, R. D. 1976. Choosing tree species for planting. pp. 193-204 In D. M. Baumgartner and R. J. Boyd (Eds.). Proc., Tree planting in the Inland Northwest. Wash. State Univ. Coop. Ext. Serv., Pullman.

165. Racey, G. D., C. Glerum, and R. E. Hutchinson. 1983. The practicality of top-root ratio in nursery stock characterization. For. Chron. 59:240-243.

166. Read, R. A. 1971. Browsing preference by jackrabbits in a ponderosa pine provenance plantation. USDA For. Serv. Res. Note RM-186. Rocky Mtn. For. and Range Exp. Sta., Ft. Collins. 4 p.

167. Ritchie, G. A. 1984. Assessing seedling quality. Chapter 23 In M. L. Duryea and T. D. Landis (Eds.). Forest nursery manual: Production of bareroot seedlings. Martinus Nijhoff/Dr Junk Publishers, The Hague, Netherlands, for Oreg. State Univ., For. Res. Lab. 340 p.

168. Ritchie, G. A. and J. R. Dunlap. 1980. Root growth potential: Its development and expression in forest tree seedlings. N.Z. J. For. Sci. 10:218-248.

169. Roberts, B. R. 1964. Effects of water stress on the translocation of photosynthetically assimilated carbon-14 in yellow poplar. pp. 272-288 In M. H. Zimmermann (Ed.). The formation of wood in forest trees. Academic Press, New York.

170. Ronco, F. 1970. Influence of high light intensity on survival of planted Engelmann spruce. For. Sci. 16:331-339.

171. Ronco, F. 1975. Diagnosis: "Sunburned trees." J. For. 73:31-35.

172. Rook, D. A. 1969. Water relations of wrenched and unwrenched Pinus radiata seedlings on being transplanted into conditions of water stress. N.Z. J. For. 14:50-58.

173. Rowe, R. N., and D. V. Beardsell. 1973. Waterlogging of fruit trees. Hort. Abstr. 43:534-548.

174. Ruehle, J. L. 1982. Mycorrhizal inoculation improves performance of container-grown pines planted on adverse sites. pp. 133-135 In R. W. Guldin and J. P. Barnett (Eds.). Proc., Southern containerized forest tree seedling conf. USDA Forest Serv. Gen. Tech. Rep. SO-37, Southern For. Exp. Sta., New Orleans.

175. Running, S. W. 1981. The influence of low soil temperatures on seedling stress. p. 67 In Proc., Exe. Sum. West. For. Conf., West. For. and Conserv. Assoc., Portland.

176. Ruth, R. H. 1968. Differential effect of solar radiation on seedling establishment under a forest stand. Ph.D. Thesis, Oreg. State Univ., Corvallis. 165 p.

177. Salisbury, F. B., and C. Ross. 1969. Plant Physiology. Wadsworth Publishing Co., Inc., Belmont, Calif. 764 p.

178. Scarratt, J. B., C. Glerum, and C. A. Plexman (Eds.). 1982. Proc., Canadian containerized tree seedling symp. Can. For. Serv., Great Lakes Res. Centre, Asulte Ste. Marie, Ontario. 460 p.

179. Scarth, G. W. 1944. Cell physiological studies of frost resistance: A review. New Phytol. 43:1-12.

180. Schmidt-Vogt, H. 1981. Morphological and physiological characteristics of planting stock. pp. 433-446 In Proc., Div. I, XVII IUFRO World Congress, Kyoto, Japan.

181. Schramm, J. R. 1958. The mechanism of frost heaving of tree seedlings. Am. Philos. Soc. Proc. 102(4):333-350.

182. Schubert, G. H. 1977. Forest regeneration of arid lands. pp. 82-87 In Proc., Soc. Amer. For. 1977 National Conv., Soc. Amer. For., Washington, D.C.

183. Schubert, G. H., and R. S. Adams. 1971. Reforestation practices for conifers in California. Div. For., State of Calif., Sacramento. 359 p.

184. Shepperd, W. D. 1981. Variation in growth of Englemann spruce seedlings under selected temperature environments. USDA For. Serv. Res. Note RM-404. Rocky Mtn. For. and Range Exp. Sta., Ft. Collins. 3 p.

185. Shirley, H. L. 1936. Lethal high temperatures for conifers and the cooling effects of transpiration. J. Agric. Res. 53:239-258.

186. Shoulders, E. 1978. Species selection in Gulf Coastal States. pp.203-210 In W. E. Balmer (Ed.). Proc., Soil moisture...site productivity symp. USDA For. Serv., Southwest Area, State and Priv. For.

187. Smith, D. M. 1962. The practice of silviculture. John Wiley and Sons, Inc., New York. 578 p.

188. Spomer, G. G. 1973. The concepts of "interaction" and "operational environment" in environmental analyses. Ecol. 54:200-204.

189. Spurr, S. H., and B. V. Barnes. 1973. Forest Ecology. The Ronald Press Company, New York. 571 p.

190. Stein, W. I. 1976. Prospects for container-grown nursery stock. pp. 89-102 In D. M. Baumgartner and R. J. Boyd (Eds.). Proc., Tree planting in the Inland Northwest. Wash. State Univ. Ext. Serv., Pullman.

191. Stein, W. I., and P. W. Owston. 1975. Why use container-grown seedlings? pp. 119-122 In Proc., Western reforestation coordinating committee, West. For. and Conserv. Assoc., Portland.

192. Stone, E. C., and R. W. Benseler. 1962. Planting ponderosa pine in the California pine region. J. For. 60:462-466.

193. Stone, E. C., and J. L. Jenkinson. 1970. Influence of soil water on root growth capacity of ponderosa pine transplants. For. Sci. 16:230-239.

194. Stone, E. C., and J. L. Jenkinson. 1971. Physiological grading of ponderosa pine nursery stock. J. For. 69:31-33.

195. Stone, E. L., R. R. Morrow, and D. S. Welch. 1954. A malady of red pine on poorly drained sites. J. For. 52:104-114.

196. Stone, E. C., and G. H. Schubert. 1959. The physiological condition of ponderosa pine (Pinus ponderosa Laws.) planting stock as it affects survival after cold storage. J. For. 57:837-841.

197. Stone, E. C., G. H. Schubert, R. W. Benseler, F. J. Baron, and S. L. Krugman. 1963. Variation in root regenerating potentials of ponderosa pine from four California nurseries. For. Sci. 9:217-225.

198. Stoszek, K. J. 1976. Protection concerns in plantation establishment. pp. 291-311 In D. M. Baumgartner and R. J. Boyd (Eds.). Tree planting in the Inland Northwest. Wash. State Univ. Coop. Ext. Serv., Pullman.

199. Strothman, R. O. 1972. Douglas-fir in northern California: Effects of shade on germination, survival, and growth. USDA For. Serv. Res. Paper PSW-84, Pac. Southwest For. and Range Exp.Sta., Berkeley. 11 p.

200. Strothman, R. O. 1980. Large stock and fertilizer improve growth of Douglas-fir planted on unstable granitic soil in northern California. USDA For. Serv. Res. Note PSW-345. Pac. Southwest For. and Range Exp. Sta., Berkeley. 7 p.

201. Tanaka, Y., J. D. Walstad, and J. E. Borrecco. 1976. The effect of wrenching on morphology and field performance of Douglas-fir and loblolly pine seedlings. Can. J. For. Res. 6:453-458.

202. Tinus, R. W., and S. E. McDonald. 1979. How to grow tree seedlings in containers in greenhouses. USDA For. Serv. Gen. Tech. Report RM-60, Rocky Mtn. For. and Range Exp. Sta., Fort Collins. 256 p.

203. Tinus, R. W., W. I. Stein, and W. E. Balmer (Eds.) 1974. Proc., North Amer. containerized forest tree seedlings symp., Great Plains Agric. Counc. Publ. 68. 458 p.

204. Toumey, J. W. , and C. F. Korstian. 1962. Foundations of silviculture. John Wiley and Sons, Inc., New York. 468 p.

205. Trappe, J. M. 1977. Selection of fungi for ectomycorrhizal inoculation in nurseries. Ann. Rev. Phytopathol. 15:203-222.

206. Trappe, J. M., and R. D. Fogel. 1977. Ecosystematic functions of mycorrhizae. pp. 205-214 In The belowground ecosystem: A synthesis of plant-associated processes. Range Sci. Dept., Sci. Ser. 26. Col. State Univ., Ft. Collins.

207. Troughton, A. 1980. Environmental effects upon root-shoot relationships. pp.25-41 In D. N. Sen (Ed.). Environment and root behavior. Goebios International, Jodhpur, India.

208. Tunner, A. 1982. A procedure for comparing alternatives in planting tree seedlings. pp. 407-418 In J. B.

224

Scarratt, C. Glerum, and C. A. Plesman (Eds.). Proc., Canadian containerized tree seedling symp. Can. For. Serv., Saulte Ste. Marie, Ontario.

209. U.S. Department of Agriculture. 1973. Silvicultural systems for the major forest types of the United States. USDA Handb. 445, Washington, D.C. 124 p.

210. van den Driessche, R. 1969. Forest nursery handbook. British Columbia For. Serv. Res. Note 48.

211. Van Eerden, E., and J. M. Kinghorn (Eds.). 1978. Proc., Root form of planted trees symp. British Columbia Min. For. and the Can. For. Serv. Joint Report 8. Victoria, British Columbia. 357 p.

212. Walker, R. B. 1954. The ecology of serpentine soils. II. Factors affecting plant growth on serpentine soils. Ecol. 35:259-266.

213. Walker, R. F., D. C. West, and S. B. McLaughlin. 1983. The development of ectomycorrhizae on containerized sweet birch and European alder seedlings for planting on low-quality sites. pp. 409-417 In E. P. Jones, Jr. (Ed.). Proc., Second biennial southern silviculture res. conf. USDA For. Serv. Gen. Tech. Rep. SE-24. Southeastern For. Exp. Sta., Asheville.

214. Waring, R. H. 1970. Matching species to site. pp. 54-61 In R. K. Hermann (Ed.). Proc., Regeneration of ponderosa pine symp. Sch. For. Paper 681, Oreg. State Univ., Corvallis.

215. Waring, R. H., and G. B. Pitman. 1983. Physiological stress in lodgepole pine as a precursor for mountain pine beetle attack. Zeitschrift fur angewandte Entomologie 96:265-270.

216. White, T. L., D. L. Lavender. K. K. Ching, and P. Hinz. 1981. First-year height growth of southwestern Oregon Douglas-fir in three test environments. Silvae Genet. 30:173-178.

217. Williamson, D. M., and D. Minore. 1978. Survival and growth of planted conifers on the Dead Indian Plateau east of Ashland, Oregon. USDA For. Serv. Res. Paper PNW-242. Pac. Northwest For. and Range Exp. Sta., Portland. 15 p.

218. Wright, J. W. 1976. Introduction to forest genetics. Academic Press, New York. 463 p.

219. Wright, J. W., W. W. Lemmien, J. N. Bright, and G. W. Parmelee. 1973. English oak - a versatile tree. Nat. Res. Res. Rep 207. Mich. State Univ., East Lansing. 6 p.

220. Zaerr, J. B. 1983. Short-term flooding and net photo-synthesis in seedlings of three conifers. For. Sci. 29:71-78.

221. Zahner, R. 1962. Terminal growth and wood formation by juvenile loblolly pine under two soil moisture regimes. For. Sci. 8:345-352.

222. Zahner, R. 1968. Water deficits and growth of trees. pp. 191-254 in T. T. Kozlowski (Ed.). Water deficits and plant growth. Vol. II. Academic Press, New York.

10. CHARACTERIZATION OF THE INTERNAL WATER RELATIONS OF LOBLOLLY PINE SEEDLINGS IN RESPONSE TO NURSERY CULTURAL TREATMENTS: IMPLICATIONS FOR REFORESTATION SUCCESS

T. C. HENNESSEY AND P. M. DOUGHERTY

Associate Professor, Department of Forestry, Oklahoma State University, Stillwater, Oklahoma 74078, and Manager, Weyerhaeuser Company Forestry Research Field Station, Wright City, Oklahoma 74766.

ABSTRACT

Considerable opportunity exists to modify the nursery environment to bring the seedling and the site closer together. Development of quantified regeneration difficulty indices based on the physical and climatic conditions associated with specific sites can provide a guide towards specifying the physiological and morphological attributes required for site adaptation, i.e., the development of target seedlings. The timing and amount of irrigation, as it influences the internal water relations of seedlings, is a particularly valuable tool in seedling architecture. Characterization of seedling water relations via the pressure-volume technique indicated that seedlings which were moderately water-stressed in late summer (rewatered when pre-dawn water potential reached -750 kPa) showed an apparent 405 kPa osmotic adjustment, a capacity for greater turgor maintenance over a range of water potentials, and significantly greater root regeneration in February as compared to unstressed seedlings. Several challenges are presented as a result of this study.

10.1 INTRODUCTION

State and industrial nurseries in the southern United States are currently producing nearly one billion pine seedlings annually (54). Presumably, the goal of nursery managers

---

Journal Article P-1590 of the Agricultural Experiment Station, Oklahoma State University, Stillwater, Oklahoma.

Duryea, M.L. and Brown, G.N. (eds.). Seedling physiology and reforestation succes
©1984, Martinus Nijhoff/Dr W. Junk Publishers, Dordrecht/Boston/London. ISBN 978-94-009-6139-5

responsible for the production of these seedlings is to produce
a seedling of acceptable quality for an acceptable price. Because plantation performance depends on the outcome of an
interaction between the stock planted and its environment,
planting stock characteristics are always a determinant of
plantation performance (9). Thus, there is a necessity for
teamwork between the nursery manager and the regeneration
forester to match seedling quality characteristics with
planting sites (16).

Numerous attempts have been made to define stock or
seedling quality based on morphological and physiological
criteria (8, 9, 12, 16, 26, 27, 40). An excellent review of
the subject was published by Ritchie (37) wherein characteristics of planting stock which reflect performance potential were
categorized as either "performance" attributes (root growth
potential, cold hardiness, stress resistance) or "material"
attributes (dormancy status, water relations, nutrition,
morphology). As most succinctly described at a 1979 IUFRO
meeting, the quality of planting stock is the degree to which
it realizes the objectives of management - "quality is fitness
for purpose". Therefore, the attributes that comprise a
"quality seedling" may vary, depending particularly on the
environmental conditions to which the seedling will be exposed
following outplanting.

This paper will emphasize that considerable opportunity
exists within the nursery to tailor seedling morphology and
physiology for better adaptability to specific sites, i.e.,
the development of target seedlings for target sites, and that
the timing and amount of irrigation, as it influences the
internal water relations of seedlings, is a particularly
valuable tool in the architecture of seedlings designed for
reforestation on specific sites.

10.2 SPECIFICATION OF SEEDLING PHYSIOLOGY AND MORPHOLOGY: DEFINING THE TARGET SEEDLING

It would seem that a critical evaluation of the factors
limiting successful regeneration on a prospective planting

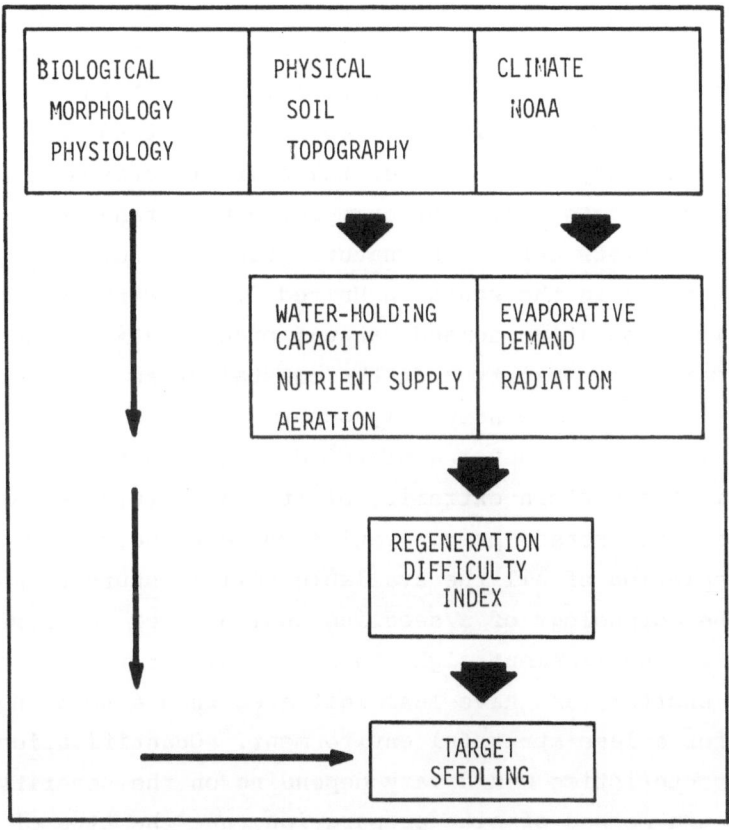

FIGURE 1. A scheme for integrating biological, physical and
climatic factors to define a target seedling.

area should be the first step towards the eventual production
of a seedling designed to survive and flourish on that site.
Often a range of site quality exists within a relatively small
geographic area. By quantification of the physical and
climatic conditions associated with a particular site, it
should be possible to develop a regeneration difficulty index
which identifies the factors limiting successful regeneration
(Figure 1). Development of this index would allow an appraisal
of the cost of altering these physical factors through various
site preparation methods. Alternatively, a seedling could be
cultured in the nursery to be both morphologically and physiologically adapted to the particular environment: in other

words, production of a seedling meeting target standards, or
a target seedling is the specification of the morphological,
anatomical, and physiological attributes of stock able to
perform under the conditions of normal use. With these sources
of information it is possible to modify stock standards or site
treatments in a way that improves the cost-effectiveness of
regeneration efforts (9). For example, a wide range of
environments exists across the natural range of loblolly pine
(Pinus taeda L.) in the southern United States with respect to
atmospheric evaporative demand, annual precipitation, and soil
moisture carrying capacity. In the coastal areas of South
Carolina, loblolly pine naturally occurs on soils where
available moisture is never depleted during the growing season.
However, in the western extremity of its range (e.g. Oklahoma),
substantial late season precipitation would be necessary to
prevent depletion of all the available soil moisture on many
sites. The morphology of a seedling designed for reforestation
in the latter environment might include a higher root to shoot
ratio, be shorter, and have less leaf area than a seedling
designed for a less stressful environment. Quantification of
these characteristics would vary depending on the severity of
the site, the method of site preparation, and the type of
planting (hand or machine). However, seedlings cultured for
reforestation in either environment should have certain
morphological characteristics in common, including a large
caliper and the potential for early root growth after outplanting (16, 26, 27, 37).

Critical differences may exist in the physiological status
of well defined target seedlings, particularly regarding factors
related to stress adaptation, i.e., traits affecting seedling
internal water relations. Identification of a seedling
targeted for reforestation on harsh sites is a long-term process
contingent on critical evaluation of planting trials and stress-
house tests to refine the needed morphological and physiological
traits.

10.3 PRODUCTION OF THE TARGET SEEDLING: THE ROLE OF GENETICS

If the target seedling has been identified for the nursery manager, the task becomes how to accomplish the production of seedlings having the desired characteristics in the most cost-efficient way possible. As a first step, it should be recognized that the nursery manager is likely dealing with many families representing widely divergent seed sources. Thus, because of genetic background, not all of the seedlings will have the same growth potential. To the nursery manager, this requires recognition that the same cultural treatments cannot be used across the nursery to develop the attributes required for seedlings targeted for different sites. As a first step, it would seem desirable to identify families in groups or blocks that have similar growth patterns in the nursery. Subsequently, the real challenge is to identify, by family, the cultural treatments needed to develop the target seedling. For some, the need may be to culture the seedlings in order to constrain their size to meet the targeted specifications. In this case, cultural treatments must be regulated to achieve the desired seedling height early in the nursery growing season followed by treatments which control height growth but allow for increased caliper and root growth with subsequent promotion of early dormancy. Thus, identification of factors such as root growth potential or chilling requirements are required for each family.

10.4 PRODUCTION OF THE TARGET SEEDLING: THE ROLE OF WATER

Traditionally, the cultural tools available to nursery managers to regulate seedling growth, or the arrows to shoot toward the target seedling, have included density regulation (7, 19, 26, 29, 43, 44, 49), undercutting and wrenching (8, 45, 50), nutrition (22, 26, 36), top pruning (26, 29, 31, 50, 54), and irrigation (4, 8, 26, 27, 29, 31, 34, 48, 55).

In the southern United States considerable work is needed towards developing an understanding of the role of water as a nursery management tool, and the interaction of irrigation and undercutting/wrenching in controlling height, diameter, and

regulation of the onset of dormancy induction. There is a need
to understand how differences in timing and amount of applied
water impact the morphology and physiology of the seedling,
and most importantly, how these changes relate to outplanting
survival and performance. The absence of this information,
particularly as related to the culture of southern <u>Pinus</u>
species, has resulted in the perpetuation of undesirable
practices by nursery managers. For example, top pruning is
routinely practiced at a majority of southern nurseries as a
method of regulating shoot-root ratios, and Williston (54)
reports that more and more nurseries are using top pruning to
produce a stockier, more uniform seedling. Use of this technique
is also endorsed for other regions of the United States (31).
In addition, many southern nursery managers irrigate heavily
even late in the growing season.

The effect of both top pruning and late season irrigation
is to delay the onset of dormancy induction. Seedlings cultured
in this fashion are less tolerant to lifting and storage
operations (34) and exhibit impaired root growth following
outplanting (27, 55) as compared to properly conditioned
seedlings. Following top pruning of loblolly pine, an apical
fascicle turns upward, leading to secondary flushing. This is
highly undesirable, because research has shown the key to
having a seedling capable of rapid root regeneration following
outplanting is culturing that seedling to be in the proper
state of dormancy prior to lifting (4, 9, 16, 20, 26, 27, 37).
For southern pines, top pruning after August requires that the
seedling be kept well-watered and therefore cultured in a rapid
growth mode if a new well-developed terminal bud is allowed to
develop. In other words, the "arrow" is being shot away from
the target. Recent research on the culture of loblolly pine
(32) has shown the impact of irrigation management on seedling
morphology. Seedlings subjected to both undercutting and
moderate moisture stress ceased height growth but caliper growth
was severely restricted. In contrast, seedlings cultured to
receive only moderate moisture stress in mid-summer continued
to increase in caliper while height growth was restricted. As

the growing season progressed, shorter photoperiod and colder temperature replaced the need for the water-stress treatment. Thus, evidence exists to demonstrate the impact of moderate water-stress on the morphology of loblolly pine seedlings. The next question is, "How is seedling physiology being impacted?"

10.5 WATER RELATIONS METHODOLOGY: A BRIEF REVIEW

Comprehensive reviews on the subject of plant water potential are available (25, 28, 46, 47, 51), therefore only a brief review is presented here. Total water potential of a plant tissue is a function of tissue water content, and is comprised of two principle components: turgor potential, the positive physical force exerted outward on the cell by the cell contents, and osmotic potential, a negative force resulting from the chemical activity of dissolved solutes held within the semipermeable membrane system of living cells (38).

While a pressure chamber measurement of plant moisture stress provides an estimate of plant water potential, it gives no information concerning the level of turgor in the cells. However, it is possible to quantify the relationship between relative water content, water potential, osmotic potential, and turgor potential of plant tissues. Because this information provides insight into a seedling's turgor maintenance mechanisms, it can be used to predict responses of different seedlings to water-related stresses, and can be used to evaluate the impact of cultural practices such as frequency and timing of irrigation on various physiological characteristics (38).

Graphically, this information is obtained by the construction of a "pressure-volume" (P-V) curve. Scholander, Hammel et al. (41) demonstrated that a pressure chamber could be used to derive a P-V curve, and Tyree and Hammel (52) later provided theoretical verification of the method. In practice, one way to obtain data to construct a P-V curve is to expose a hydrated seedling to increasing pressure in small increments in a pressure chamber, collecting the volume of sap expressed for each increment. The inverse of each balancing pressure is then plotted (ordinate) against the cumulative volume of sap expressed up to that point (abscissa).

When plotted, the curve has two distinctly different regions,
one that is curvilinear, where the chamber pressure balances
both the turgor and osmotic forces, and one that is linear,
where only the osmotic forces are balanced. Therefore, a
linear regression can be formed and extropolated to the ordinate,
giving an estimate of the reciprocal osmotic potential at full
turgor. This point is important because it establishes the
maximum turgor which can develop at full hydration: the lower
the initial osmotic potential the greater the initial turgor
pressure (38). In addition, a value can be obtained for the
osmotic potential at zero turgor. This point is important
because it establishes the lower limit of water potential at
which positive turgor can exist (11), or the point at which
"wilting" would theoretically occur. A low value would enable
a plant to maintain positive turgor while under water stress.
A more detailed discussion of the methodology can be found
elsewhere (11, 37, 38, 39).

10.6 MATERIALS AND METHODS

Research designed to investigate the relationship between
nursery water management and seedling physiology was conducted
during the 1982 growing season at the Weyerhaeuser Company
nursery in Fort Towson, Oklahoma, in cooperation with Weyerhaeuser
personnel. The objectives of the research were threefold:
(1) characterize the internal water relations of 1 + 0 loblolly
pine seedlings in response to nursery irrigation treatments,
(2) determine if culturally induced changes in seedling
physiology were transient or permanent, and (3) measure the
impact of nursery irrigation treatments on seedling root
regeneration potential. The information gained from this study
was expected to assist the nursery manager in refining techniques
for the development of target seedlings.

The experimental design utilized one loblolly pine family
from North Carolina. Seedlings were operationally cultured
until August 23, at which time two water regimes were implemented.
One portion of the seedlings was maintained in a well-watered

condition, i.e., rewatered whenever seedling xylem pressure potential reached a pre-dawn level of -250 kPa as measured by a pressure chamber (Model 3005, Soil Moisture Equipment Company, Santa Barbara, CA). Seedlings growing in a separate part of the seedbed were subjected to a moderate level of moisture stress, i.e., rewatered only when seedling xylem pressure potential reached a pre-dawn level of -750 kPa. Irrigation treatments were maintained until October 28, when the reoccurrence of fall precipitation negated the stress treatment. During September only 1.70 mm of precipitation was recorded, thus allowing well-defined treatment differences to be imposed. Replicated seedlings from each treatment were sampled and a composite P-V curve was constructed for each treatment on August 27, October 1, and November 5, 1982 and January 8, 1983.

On February 8, 1983 seedlings were lifted from the nursery, and a controlled environment growth chamber study was initiated. A total of 144 seedlings (72 from each irrigation treatment) were individually planted into 1.9 l containers containing a very fine sandy loam soil (pH 6.7). Temperature was regulated to provide an average day/night soil temperature of 9°C/7°C with a 16 hour photoperiod. Seedlings were well-watered throughout the growth chamber experiment. To monitor root growth potential, a subsample of the seedlings was excavated on days 15, 25, and 35 following the initiation of the experiment, and the number of new white roots longer than one centimeter were counted.

10.7 RESULTS AND DISCUSSION

Data for the measured bulk water relations parameters by month and irrigation treatment are shown in Table 1. Results show that on August 27, following only four days of the irrigation treatments, values of the osmotic potential at both full and zero turgor were lower for the well-watered seedlings. However, measurements taken in subsequent months consistently showed that seedlings that had been moderately water-stressed starting in August had lower values for both water relations parameters than seedlings which had received the well-watered treatment.

Table 1. Bulk water relations parameters for 1 + 0 loblolly
pine seedlings by date and irrigation treatment. Stress
treatment initiated on August 23, 1983.

Date	Irrigation treatment[1]	Osmotic potential at full turgor (kPa)[2]	Osmotic potential at zero turgor (kPa)
Aug. 27	W	−1431 a[3]	−3225
	S	−1300 a	−2500
Oct. 1	W	−1169 a	−2703
	S	−1284 a	−3125
Nov. 5	W	−1175 a	−2564
	S	−1423 a	−3225
Jan. 8	W	− 891 a	−2778
	S	−1296 b	−3125

[1]W = well-watered; S = water-stressed

[2]100 kPa = 1 bar

[3]Treatment means at each date followed by the same letter do
not differ significantly at P = 0.05

In January, over four months after the initiation of the
irrigation treatments, the osmotic potential at full turgor
for seedlings that had been moderately water-stressed was
405 kPa lower than the value for the well-watered seedlings.

The development of lower values of the osmotic potential
at full and zero turgor in those seedlings which were
moderately water-stressed suggests those seedlings may have
undergone an osmotic adjustment, i.e., the accumulation of
solutes in sufficient quantity to change the osmotic potential.
Boyer (5) indicated that the high concentration of solutes
found in plant cells creates the ultimate driving force, or
osmotic potential, that brings water into the plant from the
soil. Within a particular tissue, the osmotic potential
depends on the balance between rate of solute accumulation
and rate of use by the cells. Meyer and Boyer (33) provided

evidence that osmotic adjustment could occur in response to dry soil conditions. Similar findings (2, 13, 14, 17, 23, 24, 35, 42) confirmed this result with a range of species. The benefit of this process is that turgor can be sustained and plant growth maintained on limited soil water when growth otherwise would not occur (5). Thus, the ability of a seedling to osmotically adjust in response to water-stress may be an important survival mechanism. An osmotic adjustment has been reported to allow the maintenance of cell elongation (33) and stomatal opening and photosynthesis (3, 18), as well as enhancing tolerance to dehydration and promoting exploration of greater soil volume for water through root growth (24).

Agronomists have made considerable practical use of information concerning osmotic adjustment. For example, it is reported that in Australia varieties of wheat selected for the ability to osmotically adjust outyielded commercial cultivars by almost 2:1 under dry conditions (5). It is possible that foresters will be able to develop similar screening indices to select families of loblolly pine for reforestation on droughty sites based on the ability to osmoregulate. Currently, no information is available concerning the degree of genetic variation which may exist for this trait.

The P-V technique also provides a means of evaluating the capacity of a plant to maintain high turgor at decreasing water potential, or turgor maintenance capacity. Because plant growth is a turgor-dependent process, a plant's ability to maintain positive cell turgor over a wide range of water potentials is a key adaptation to water-related stress (38). Data for the turgor component of water potential, by date and irrigation treatment, are shown in Table 2. Well-watered seedlings sampled on August 27 had higher turgor potential at any given level of water potential compared to water-stressed seedlings. However, as the season progressed this changed, so that by January 8 the turgor potential for the moderately water-stressed seedlings at a water potential value of -724 kPa, for example, was nearly twice that of seedlings which had been well-watered in August and September.

Table 2. A comparison of seedling turgor potential at various values of water potential for well-watered and water-stressed seedlings.[1]

Date	Water potential	Nursery Irrigation Treatment	
		Well-watered turgor potential	Water-stressed[2] turgor potential
Aug. 27	- 741	872	664
	-1266	586	424
	-1852	370	248
	-2174	258	181
Jan. 8	- 724	452	899
	-1019	451	774
	-1586	377	515
	-2127	254	410

[1]All units kPa
[2]Stress treatment initiated on August 23

The maintenance of turgor pressure as the plant water potential declines is crucial for cell expansion, for growth and for many of the associated biochemical, physiological and morphological processes (21). The ability of the water-stressed seedlings to maintain higher turgor in January, typically a time of outplanting in the southern United States, could be important in promoting early survival and growth. The importance of rapid establishment was shown by McCluskin (30), who reported that first year survival of loblolly pine seedlings was dramatically reduced by early season drought.

Table 3 shows mean values of periodic root regeneration for seedlings which had been either well-watered or moderately water-stressed for approximately two months while in the nursery during the previous growing season. After 35 days exposure to relatively cold soil temperatures, those seedlings which had

been moderately water-stressed in the nursery had initiated
nearly three times as many new roots as unstressed seedlings.
The ability of the water-stressed seedlings to more rapidly
initiate new roots in a cold environment, if true for field
plantings, would provide for more rapid establishment following
outplanting and presumably allow those seedlings to be more
competitive for available soil water during times of seasonal
moisture stress.

Table 3. Effect of nursery irrigation treatment on periodic
root regeneration under controlled environment conditions.[1,2]

| Harvest day | Nursery Irrigation Treatment | | LSD |
	Root regeneration well-watered	Root regeneration water-stressed	
15	0.2	0.5	0.62
25	1.5	3.6	2.44
35	2.4	7.1	1.75

[1]Average day/night soil temperature $9^{o}C/7^{o}C$. All seedlings
well-watered while in growth chamber.

[2]Means in the same column connected by a vertical line do not
differ significantly at $P = 0.05$.

The importance of vigorous root growth by outplanted seedlings
as soon as the environmental conditions permit has been emphasized by others (8, 9, 16, 26, 27, 37). Burdett (9) placed high
root growth capacity at the first of a list of stock characteristics which enhance early plantation performances, and
Bunting (8) indicated that root regeneration is one of the
single most important physiological attributes the seedling can
have.

Phenological observations made throughout the course of this study confirm that the inducement of a moderate level of moisture stress within loblolly pine seedlings during late August and September is an effective method for controlling seedling height expansion while simultaneously allowing for increased caliper growth and the development of seedlings that will have improved tolerance to water stress. It is reasonable to speculate that planting seedlings cultured in this manner could dramatically improve reforestation success over a wide range of sites in the southern United States. Ursic (53) stated that moisture stress imposed by drought, adverse sites, and competing vegetation largely determine the early performance of outplanted southern pines. Further, Cannel et. al. (10) found internal seedling water stress to be the most critical factor limiting height and volume growth of loblolly pine during the growing season, even on wet sites, in the southern United States.

10.8 SUMMARY AND RECOMMENDATIONS

This paper has emphasized that considerable opportunity exists to modify the nursery environment to bring the seedling and the site closer together. Development of quantified regeneration difficulty indices based on the physical and climatic conditions associated with specific sites can provide a guide towards specifying the physiological and morphological attributes required for site adaptation, i.e., the development of target seedlings. The time and effort expended in the culture of target seedlings to enhance regeneration success can be measured against the costs associated with replanting, additional site preparation or lost yield.

Results of this research indicate that the timing and amount of irrigation, as it influences the internal water relations of seedlings, can be a valuable tool in developing seedlings designed for reforestation on specific sites. In this study, an irrigation regime which cycled around a pre-dawn water potential of -750 kPa for approximately two months induced changes in both the osmotic and turgor potential of 1 + 0 loblolly pine seedlings. These changes were still evident in January. Moderately water-stressed seedlings showed an apparent 405 kPa osmotic adjustment and a capacity for greater turgor maintenance over a range of water potentials. These characteristics would be expected to enhance reforestation success particularly in areas subject to summer droughts. In addition, seedlings which were moderately water-stressed in the nursery were subsequently found to have significantly greater root regeneration in cold soil as compared to well-watered seedlings after 35 days.

Finally, observations of seedling phenology indicated that a moderate degree of water-stress was sufficient to control height growth of loblolly seedlings while allowing for increased seedling caliper and the normal sequencing of dormancy induction. In short, nursery water management appears to be a valuable "arrow" when shooting at the development of a target southern pine seedling.

As a result of these experiments, several challenges appear. First is the need to relate changes in seedling morphology and physiology, induced by various irrigation regimes, to field survival and performance. The effect of an apparent 405 kPa osmotic adjustment and greater turgor maintenance capacity as it relates to tolerance following outplanting is unknown. The second challenge is to evaluate genetic differences between families in the capacity for culturally induced changes in seedling physiology and morphology, and to investigate further the underlying response mechanisms. Elucidation of these mechanisms may make it possible to better tailor nursery cultural practices to improve seedling drought resistance, frost hardiness or root regeneration potential. Thirdly, there is a need to better understand the interaction of seedling water-stress and other nursery practices (e.g., undercutting) in the culture of target seedlings.

A review of the literature describing current nursery practices indicates that these challenges are far from being met. In a 1977 survey of 134 public and industrial nurseries in the United States to define research needs in forest nursery practices, there was no mention of the need to better understand

the role of irrigation as a nursery management tool (1).
Another report (31) indicated that the pressure chamber is
only used in about 10-15 percent of western United States
nurseries to monitor irrigation. Finally, a recent survey of
60 state, federal, and industrial nurseries in the southern
United States reported that none monitor xylem water potential
operationally during any phase of seedling culture (6). Clearly,
there is a long way to go to meet Day's (15) recommendation
that in the 1980's, all forest tree nurseries should be
implementing scientifically planned irrigation programs designed
to optimize growth and the morphological and physiological
quality of nursery stock.

REFERENCES

1. Abbott, H. G., and S. D. Fitch. 1977. Forest nursery
    practices in the United States. Journal of Forestry 75:
    141-145.
2. Acevedo, E., E. Fereres, T. C. Hsiao, and D. W. Henderson.
    1979. Diurnal growth trends, water potential, and osmotic
    adjustment of maize and sorghum leaves in the field. Plant
    Physiol. 64: 476-480.
3. Biscoe, P. V. 1972. The diffusion resistance and water
    status of leaves of Beta vulgaris. Journal of Experimental
    Botany 23: 930-940.
4. Blake, J., J. Zaerr, and S. Hee. 1979. Controlled moisture
    stress to improve cold hardiness and morphology of
    Douglas-fir seedlings. Forest Science 25: 576-582.
5. Boyer, J. 1983. Subcellular mechanisms of plant response
    to low water potential. In: J. F. Stone and W. O. Willis,
    eds., Plant Production and Management Under Drought
    Conditions. Elsevier Science Publishers B. V., 1983.
6. Boyer, J. N., and D. B. South. 1984. Forest nursery
    practices in the South. Southern Journal of Applied
    Forestry (in press).
7. Bunting, W. 1973. Seedbed density trials: white pine,
    red pine, and white spruce. Ontario Ministry of Natural
    Resources, Nursery Note No. 34. 17 p.
8. Bunting, W. R. 1980. Seedling quality: growth and
    development. Pages 21-42. In: Proc. N. Amer. Forest
    Tree Nursery Soils Workshop. Syracuse, New York.
    USDA/CFS sponsored publication.
9. Burdett, A. N. 1983. Quality control in the production
    of forest planting stock. The Forestry Chronicle 59:
    132-138.

10. Cannel, M., F. Bridgewater, and M. Greenwood. 1978. Seedling growth rates, water stress responses, and root-shoot relationships related to eight-year volumes among families of Pinus taeda L. Silvae Genetica 27: 237-248.

11. Cheung, Y. N. S., M. Tyree, and J. Dainty. 1975. Water relations parameters on single leaves obtained in a pressure bomb and some ecological interpretations. Canadian Journal of Botany 53: 1342-1346.

12. Cleary, B. D., R. D. Graves, and P. W. Owston. 1978. Seedlings. Pages 63-97. In: Regenerating Oregon's Forests (B. D. Cleary, R. D. Graves, and R. K. Hermann, eds.). Oregon State University, Extension Service, Corvallis, Oregon.

13. Cutler, J. M., D. W. Rains, and R. S. Loomis. 1977. Role of changes in solute concentration in maintaining favorable water balance in field-grown cotton. Agron. J. 69: 773-779.

14. Cutler, J. M., K. W. Shahan, and P. L. Steponkus. 1980. Influence of water deficits and osmotic adjustment on leaf elongation in rice. Crop Sci. 20: 314-318.

15. Day, R. 1980. Effective nursery irrigation depends on regulation of soil moisture and aeration. Pages 52-71. In: Proc. N. Amer. Forest Tree Nursery Soils Workshop. Syracuse, New York.

16. Edgren, J. 1980. The reforestation system - a team effort. Pages 12-20. In: Proc. N. Amer. Forest Tree Nursery Soils Workshop. Syracuse, New York.

17. Fereres, E., E. Acevedo, D. W. Henderson, and T. C. Hsiao. 1978. Seasonal changes in water potential and turgor maintenance in sorghum and maize under water stress. Physiol. Plant. 44: 261-267.

18. Goode, J. E., and K. H. Higgs. 1973. Water, osmotic, and pressure potential relationships in apple leaves. Journal of Horticulture Science 48: 203-215.

19. Harms, W. R., and O. Langdon. 1977. Competition-density effects in a loblolly pine seedling stand. United States Forest Service Research Paper SE 161. 8 pgs.

20. Hermann, R. K. 1967. Seasonal variation in sensitivity of Douglas-fir seedlings to exposure to roots. Forest Science 13: 140-149.

21. Hsiao, T. C. 1973. Plant responses to water stress. Annual Review of Plant Physiology 24: 519-570.

22. Ingestad, T. 1979. Mineral nutrient requirements of Pinus sylvestris and Picea abies seedlings. Physiologia Plantarum 45: 373-380.

23. Jones, M. M., C. B. Osmond, and N. C. Turner. 1980. Accumulation of solutes in leaves of sorghum and sunflower in response to water deficits. Aust. J. Plant Physiol. 7: 193-205.

24. Jones, M. M., and N. C. Turner. 1978. Osmotic adjustment in leaves of sorghum in response to water deficits. Plant Physiol. 61: 122-126.

25. Jones, M. M., N. C. Turner, and C. B. Osmond. 1980. The physiology and biochemistry of drought resistance. L. Paleg and D. Aspinall, eds. Academic Press, N. Y.

26. Lavender, D. 1984. Plant physiology and nursery environment: interactions affecting seedling growth. M. Duryea and T. Landis, eds. In: Forest Nursery Manual: Production of Bareroot Seedlings. Martinus Nijhoff/Dr W. Junk Publishers, The Hague-Boston-London (in press).

27. Lavender, D. and B. Cleary. 1974. Coniferous seedling production techniques to improve seedling establishment. In: Proc. N. Amer. Containerized Forest Tree Seedling Symposium, Denver, Colorado. Pages 177-180.

28. Levitt, J. 1980. Responses of plants to environmental stresses. Vol II. Water, radiation, salt, and other stresses. Academic Press, N. Y. 607 p.

29. Lopushinsky, W. and T. Beebe. 1976. Relationship of shoot-root ratio to survival and growth of outplanted Douglas-fir and ponderosa pine seedlings. USFS, Pacific NW Forest and Range Experiment Station, Portland, OR. Res. Note PNW-274. 7 p.

30. McCluskin, D. 1966. Survival of planted loblolly pine seedlings: moisture, temperature, and soil influences. Journal of Forestry 64: 731-734.

31. McDonald, S. and S. Running. 1979. Monitoring irrigation in western forest nurseries. USFS Rocky Mnt. For. and Range Expt. Station General Tech. Rep. RM-61. 8 p.

32. Mexal, J. Date unknown. Unpublished data. Weyerhaeuser Co., Hot Springs, Ark.

33. Meyer, R. F. and J. S. Boyer. 1972. Sensitivity of cell division and cell elongation to low water potentials in soybean hypocotyls. Planta 108: 77-87.

34. Morby, F. E. 1981. Irrigation regimes in a bareroot nursery. Pages 55-59. In: Proc. 1981 Intermountain Nurserymen's Assoc. Mtg., August 11-13, Edmonton, Alberta.

35. Morgan, J. M. 1977. Differences in osmoregulation between wheat genotypes. Nature 270: 234-235.

36. Radwan, M., G. Crouch, and H. Ward. 1971. Nursery fertilization of Douglas-fir seedlings with different forms of nitrogen. U.S. Forest Service, Pacific NW Forest and Range Expt. Station, Portland, OR. Res. Paper PNW-113. 8 p.

37. Ritchie, G. 1984. Assessing seedling quality. M. Duryea and T. Landis, eds. In: Forest Nursery Manual: Production of Bareroot Seedlings. Martinus Nijhoff/Dr W. Junk Publishers, The Hague-Boston-London (in press).

38. Ritchie, G. and D. Dunham. 1979. P-V Curves I. Theory, Methods, and Interpretation. Weyerhaeuser For. Res. Tech. Rep. 042-2203/79/42. 16 p.

39. Ritchie, G. and T. Hinkley. 1975. The pressure chamber as an instrument of ecological research. Advances Ecol. Res. 9: 165-254.

40. Schmidt-Vogt, H. 1981. Morphological and physiological characteristics of planting stock: present state of research and research tasks for the future. Pages 433-446. In: Proc. IUFRO XVII World Congress, Kyoto, Japan.
41. Scholander, P. F., H. T. Hammel, E. D. Bradstreet, and E. A. Hemmingsen. 1965. Sap pressure in vascular plants. Science 148: 339-46.
42. Sharp, R. E. and W. J. Davies. 1979. Solute regulation and growth by roots and shoots of water-stressed maize plants. Planta 147: 43-49.
43. Shipman, R. D. 1966. Low seedbed densities can improve early height growth of planted slash and loblolly pine seedlings. Tree Planter's Notes 76: 24-29.
44. Shoulders, E. 1961. Effect of nursery bed density on loblolly and slash pine seedlings. Journal of Forestry 59: 576-579.
45. Shoulders, E. 1963. Root-pruning southern pines in the nursery. United States Forest Service Research Paper SO-5. 6 pgs.
46. Slavik, B. 1974. Methods of studying plant water relations. Springer-Verlag., N. Y. 449 p.
47. Slayter, R. O. 1967. Plant-water relationships. Academic Press, N. Y. 336 p.
48. Stransky, J. and D. Wilson. 1964. Terminal elongation of loblolly and shortleaf pine seedlings under soil moisture stress. Pages 439-440. In: Proc. Soil Science Society, Vol 28.
49. Switzer, G. and L. Nelson. 1963. Effects of nursery fertility and density on seedling characteristics, yield, and field performance of loblolly pine (Pinus taeda L.). Proc. Soil Science Society of America 27: 461-64.
50. Tanaka, V., J. Walstad, and J. Borrecco. 1976. The effect of wrenching on the morphology and field performance of Douglas-fir and loblolly pine seedlings. Canadian Journal of Forestry Research 6: 453-458.
51. Turner, N. C. and P. J. Kramer. 1980. Adaptation of plants to water and high temperature stress. John Wiley and Sons, N. Y. 482 p.
52. Tyree, M. T. and H. T. Hammel. 1972. The measurement of the turgor pressure and the water relations of plants by the pressure-bomb technique. Journal of Experimental Botany 23: 267-282.
53. Ursic, S. 1961. Tolerance of loblolly pine seedlings to soil moisture stress. Ecology 42: 823-825.
54. Williston, H. L. 1980. What the southern tree planter wants from the nurseryman. Pages 6-11. In: Proc. N. Amer. Forest Tree Nursery Soils Workshop. Syracuse, N. Y. USDA/CFS sponsored publication.
55. Zaerr, J. B., B. D. Cleary, and J. Jenkinson. 1980. Scheduling irrigation to induce dormancy. Pages 74-78. In: Proc. Intermountain Nurserymen's Association and Western Forest Nursery Association Meeting, Boise, Idaho.

Planting Site and Stock Response

Accelerating Early Growth in Plantations

- **Vegetation management**—Papers 11 and 12
- **Nutrition management**—Papers 13 and 14

11. GROWTH RESPONSE AND PHYSIOLOGY OF TREE SEEDLINGS AS
AFFECTED BY WEED CONTROL

D. H. GJERSTAD, L. R. NELSON, J. H. DUKES, JR., AND
W. A. RETZLAFF
Associate Professor, Research Associate, Graduate Research
Assistant, Graduate Research Assistant, Department of
Forestry, Auburn University, Auburn, Alabama 36849

ABSTRACT

Tree growth is dependent on the availability of light,
water, nutrients, CO_2 and O_2. Unless trees are overtopped
by vegetation, CO_2, O_2 and sunlight are usually not limiting
factors. However, intraspecific competition often results
in a limited supply of water and nutrients to crop trees.
Such competition can result in lower survival in new
plantings and reduced growth in surviving stock.

Recent forest vegetation management research in
combination with several older studies demonstrate large
survival and growth gains in newly established forests.
Such responses have resulted from reductions in woody or
herbaceous or both types of competition. Only in recent
years have researchers attempted to relate the effects of
vegetation management to tree physiology. It appears that
in most cases competition results in water being the
limiting factor for tree survival and growth.

11.1. INTRODUCTION

Weed control in agriculture was likely initiated by
prehistoric man to enhance the survival and yield of food
crops. Although control techniques have evolved from using
hands and sticks to mechanical and chemical methods, the
basic objective to maximize yields has not changed.
Likewise as productive land available to forestry declines
and demands for forest products increase, foresters are

Duryea, M.L. and Brown, G.N. (eds.). Seedling physiology and reforestation succes
©1984, Martinus Nijhoff/Dr W. Junk Publishers, Dordrecht/Boston/London. ISBN 978-94-009-6139-5

implementing cultural techniques to increase yields. As in
agronomic crops, weed control in forestry has tremendous
potential to boost productivity.

Until recent years, intensive weed control in forestry
has been limited to high valued seed orchard and nursery
crops where competition obviously reduced yield and quality.
However foresters are becoming more aware that competition
for light, water and nutrients by undesired species is
causing growth losses in commercial tree species (40). A
survey on forest rehabilitation conducted in 1973 indicated
some 300 million acres of commercial forest land in need of
stand improvement or conversion to merchantable species
(39).

This paper will discuss seedling response when competing
vegetation is controlled and seedling physiological factors
influenced by such control.

11.2 GROWTH RESPONSE AND SURVIVAL FOLLOWING WEED CONTROL

Several studies demonstrate increased growth rates and
survival of outplanted tree seedlings following herbaceous
weed control. Even short-term results (1 or 2 growing
seasons) have shown dramatic seedling height and diameter
responses (13, 17, 25, 26). Nambiar and Zed (25) noted a
10-140% increase in stem volume and 40% less mortality one
year following herbaceous weed control in Pinus radiata (D.
Don), while Nelson et al. (26) found as much as a twelve
fold increase in loblolly (P. taeda L.) biomass one year
after outplanting on weeded plots. Although no survival
differences were noted, Knowe et al. (20) observed that
four-year-old outplanted loblolly pine seedlings following
100% weed control for the initial two years were twice the
height and diameter as seedlings with no weed control.

In a similar study involving longleaf pine (P. palustris
Mill.), trees on weeded plots had at least one year's greater
height growth at age 3 with much of the differences
attributed to a larger proportion of seedlings being out of
the grass stage (43). It is well documented that artificial

regeneration of longleaf pine has been limited due to
delayed growth resulting from herbaceous weed competition
(5, 27, 28). With the advent of effective herbicide
treatments, longleaf pine will likely gain new popularity
for outplanting.

Although short-term growth response studies have
stimulated interest in the potential of weed control, long-
term observations are needed to determine whether initial
differences are maintained through a rotation. Results from
a limited number of studies indicate that differences in
volume continue through the initial 20 years of a rotation.
Twenty years after releasing grass-stage longleaf pine from
hardwood competition, Michael (23) found a 40% volume or eight-
year growth advantage for treated plots. Nineteen years
following release of ponderosa pine (P. ponderosa Laws.)
from bearmat (Chamaebatia foliolosa Benth.), survival and
height growth were 90% and 250% greater respectively on the
best weed control treatment as compared to the untreated
(36). Ten-year-old radiata pine on areas with grass control
had a height advantage equivalent to two years growth and were
50% larger in diameter (1).

11.3 MOISTURE STRESS AND SEEDLING RESPONSE

Similar growth differences and trends have been noted in
moisture stress treatments as in weed control studies.
Zahner (41) examined seedlings planted in drums with wet and
dry treatments and found that seedlings in the dry treatment
stopped growing in mid-July whereas the wet continued
growing into October. It was also noted that the watered
seedlings put out four flushes of height growth as compared
to two in the dryer treatment and that the first flush in
the following year was 25% longer for the wet treatment.

Water stress was observed by Cannell et al. (8) to
severely depress first year loblolly pine seedling growth
and high correlations between growth and soil moisture were
only found when soil moisture was limiting. Stransky and
Wilson (35) found with 1-0 potted seedlings that height

growth was inhibited at soil moisture tensions no greater
than -2 atm and stopped at -3.5 atm. Mature stand loblolly
pine diameter growth has also been related to soil moisture
(2) and more specifically latewood production is highly
correlated with moisture (6).

Soil moisture stress during or prior to a certain
growing period affects bud set and resulting vegetative
growth of the plant. Clements (9) found that red pine (P.
resinosa Ait.) irrigated frequently for one year had larger
buds and more resulting growth the following year than did
buds in a drought treatment. Similarly loblolly and white
pine bud and needle growth decreased as mean soil water
potential decreased (19). Loblolly pine, irrigated under
two soil moisture regimes, produced different amounts of
growth (41). Trees grown under wet conditions developed four
internodes per year, while only two formed under dry
conditions. The influence of the previous season
environmental conditions was demonstrated by a 25% longer
initial flush for wet trees.

Root development and growth is dependent on the
availability of soil moisture. Root growth of loblolly pine
and white pine (P. strobus L.) decreased as mean soil water
potential decreased (19). Root regeneration of white pine
seedlings is also delayed at increasing soil moisture
tensions (11). Northern red oak (Quercus rubra L.) root
regeneration and increasing root length were inhibited at -4
bars and then stopped completely at -6 bars (21).

In a study by Pessin (27), longleaf and slash pine (P.
elliottii Engelm.) seedlings were grown under varying
moisture conditions ranging from very dry (5% soil moisture
content, SMC) to wet conditions (35% SMC). Taproots of longleaf seedlings in wet soil were longer than in any other
culture, but numbers of lateral roots were least. Slash
pine taproots were the shortest in both very dry and wet
soils and longest in damp and moist soils. However root
growth is affected less than shoot growth by varying soil
moisture content. Hatzistathis (16) found that root growth

of loblolly and shortleaf pine (P. echinata Mill.) was
affected less than shoot growth under increasing moisture
stress.

11.4 PHYSIOLOGICAL RESPONSE

Seedling physiological response to competition and
moisture stress has been measured by leaf moisture content,
transpiration, diffusive resistance, photosynthesis and
respiration. Nelson et al. (26) found that pre-dawn needle
xylem tensions were significantly different between weeded
and nonweeded plots with significant seedling growth
differences.

The effect of needle moisture stress on the physiology
of the needle has generally been observed to decrease
transpiration, close stomates, reduce photosynthesis, and
have variable effects on respiration. Second year loblolly
pine seedlings were found to decrease carbon dioxide uptake
with increasing diffusion pressure deficit of the needles
and photosynthesis was found to be highly correlated
(positive) with transpiration rate (4). It was concluded
that stomatal aperture and mesophyll resistance both could
play a part in photosynthesis reductions but the former was
most likely dominant. This conclusion is supported by both
Miller and Allen (24) and Bilan et al. (3). Using detached
needles in a Warburg respirometer, Miller and Allen (24) found
high positive correlations between both photosynthesis and
stomatal aperture, and relative needle water content and
stomatal aperture. Bilan et al. (3) studying the drought
resistant "Lost Pines" loblolly pine variety of east Texas
found that with water stress this variety had fewer percent
open stomates and hence a greater reduction in transpiration
than other East Texas loblolly pine seedlings. Percent open
stomates was highly correlated (positive) with transpiration
rate over a range of moisture conditions and was also highly
correlated with needle water content but only in low soil
moisture situations. At high soil moisture treatments,
transpiration was highly correlated with needle water

content but not at low soil moistures. These data suggest
that transpiration may be limited by mesophyll resistances
in high moisture situations, but under moisture stress
conditions stomatal control is dominant.

In contrasting findings, Tolley and Strain (37) found
that atmospheric carbon dioxide enrichment with both
stressed and well watered potted seedlings increased
photosynthesis little. As stomatal resistance was found not
to be the limiting factor, it was concluded that mesophyll
resistance restricted increased CO_2 uptake.

Seedling physiology appears to be very sensitive to
internal moisture levels since leaf moisture content changes
very little before seedling growth stops (35). This is the
basis for the general observation that most plant growth
occurs at night when plant moisture stress is less.

Needle moisture measures have produced varied results in
predicting seedling mortality. Seedling death was found by
Ursic (38) to be predictable by needle moisture content with
80% water as a percentage of dry weight being the lethal
limit. Brix (4), on the other hand, found 110% needle
moisture content to be the lethal threshold. He concluded
that a leaf DPD of 27 atm was the limit. Stransky (34)
preferred a range of needle moisture contents after
observing that at moisture contents from 105-65% a seedling
could live or die while at 85% it had a 50% chance of
survival.

Dark respiration of loblolly pine needles with
increasing water stress was noted by Brix (4) to first
decrease with increasing leaf DPD and then to increase to
140% of that observed with seedlings at field capacity and
finally to decrease as leaf DPD increased further. He
summarized that possibly this response was due to an initial
decrease in available sugars due to the concomitant decrease
in photosynthesis followed by a temporary increase due to
starch hydrolysis.

Photosynthesis is less sensitive than most physiological responses to moisture stress. Photosynthesis maintained 100% of maximum until leaf DPD was elevated above 7 atm and became zero at 14 atm (4). Recovery was slow with 100% recovery of net photosynthesis taking 50 hours after rewatering the dry seedlings. Of interest was that recovery was hastened if the seedling was cut from its rootstock and placed in water. Evidently root water uptake resistance is a limiting factor.

Both McGregor and Kramer (22), and Drew and Ledig (12) found net photosynthesis to be lowest in winter and peaking in late spring and summer. They concluded that the initial peak and later decline in photosynthesis were due to increasing photosynthetic capacity of the needles. Drew and Ledig (12) observed a net photosynthesis lag in June that corresponded with increasing shoot respiration and growth indicating an accelerated photosynthate use rather than increased photosynthetic capacity.

On a whole seedling basis, photosynthesis peaked in September (22). Burkhalter et al. (7) found negative correlations between total seedling volume, and net photosynthesis, total photosynthesis and dark respiration. Mutual leaf shading was believed to be the cause. Leaf chlorophyll content has been correlated with photosynthetic capacity, but both Burkhalter et al. (7) and McGregor and Kramer (22) found this not to be the case with nonstressed loblolly pine seedlings.

11.5 RESEARCH NEEDS

Little research has specifically been designed to examine seedling response following weed control. Ample opportunities exist in most ecosystems to examine ecological, physiological and growth response following vegetation management. With the advent of effective weed control techniques, researchers are now in the position to examine the effects of intra- and interspecific competition on seedling physiology and growth (10, 36, 42). Scant

knowledge exists on the degree of interference by various
plant species in the survival and growth of tree seedlings.
Ecological studies are needed that define the competitive
relationships between species.

To better understand the fundamental relationship
between environmental factors and seedling response,
research is needed in defining physiological parameters
involving tree growth limitations. As was emphasized in
this paper, moisture and nutrients appear to be the limiting
environmental factors on most sites for seedling survival
and growth. However on specific sites the limiting factors
may include genetic material, shading, temperature extremes,
light intensity or allelopathy (14, 18, 30, 43). Although
allelopathy attracts considerable interest, only limited
effort has gone into determining allelopathic effects of
plants on forest trees. Allelopathic research has been
limited to observing tree growth which has been impeded and
relating this reduced growth rate to nearby plants.

Some are concerned that rapid tree growth following weed
control may change wood properties. However several studies
comprising cultural practices which increased growth rates
have concluded that wood properties such as specific
gravity, fiber length, fibril angle and fiber dimensions are
related to the age of wood from the pith, not growth rate
(15, 31, 32, 33). Research specifically involving weed
control is needed to verify these findings.

It should not be concluded that weed control will solve
all regeneration problems, as it is only one of several
cultural considerations necessary for the successful
regeneration of tree seedlings. Artificial regeneration is
a system involving genetics, nursery management, planting
practices and consideration of environmental conditions on
the planting site. If any component within the system
fails, seedling survival and growth will be diminished.
While time, space, and funding constraints limit variables in
a given study, research incorporating as many components of
the system as possible is needed. When this is not

possible, the components should be described to allow more accurate comparisons between studies.

REFERENCES

1. Balneaves, J. M. 1982. Grass control. In What's new in forest research. Forest Research Institute, Rotorua, NZ. No. 113. 4 p.
2. Bassett, J. R. 1972. Tree growth as affected by soil moisture availability. Soil Sci. Soc. Proc. 28:436-438.
3. Bilan, M. V., C. T. Hogan, and H. B. Carter. 1977. Stomatal opening, transpiration, and needle moisture in loblolly pine seedlings from two Texas seed sources. For. Sci. 23:457-462.
4. Brix, H. 1962. The effect of water stress on the rates of photosynthesis and respiration in tomato plants and loblolly pine seedlings. Physiol. Plant. 15:10-20.
5. Bruce, D. 1959. Effect of low competition on longleaf pine seedling growth. Soc. Amer. For. Proc. 1958:151-153.
6. Buckingham, F. M. and F. W. Woods. 1969. Loblolly pine (Pinus taeda L.) as influenced by soil moisture and other environmental factors. J. Appl. Ecol. 6(1):47-59.
7. Burkhalter, A. P., C. F. Robertson, and M. Reines. 1967. Variation in photosynthesis and respiration in southern pines. Ga. For. Res. Paper No. 46.
8. Cannell, M. G. R., F. E. Bridgwater, and M. S. Green. 1978. Seedling growth rates, water stress responses and root shoot relationships related to eight-year volumes among families of Pinus taeda. Silvae Genetica. 27(6):237-248.
9. Clements, J. R. 1970. Shoot responses of young red pine to watering applied over two seasons. Can. J. Bot. 48:75-80.
10. Conard, S. G. and S. R. Radosevich. 1981. Photosynthesis, xylem pressure potential, and leaf conductance of three montane chaparral species in California. For. Sci. 27(4):627-639.
11. Day, R. J. and G. R. MacGillivray. 1975. Root regeneration of fall-lifted white spruce nursery stock in relation to soil moisture content. For. Chron. 51:196-199.
12. Drew, A. P. and F. T. Ledig. 1981. Seasonal patterns of CO_2 exchange in the shoot and root of loblolly pine seedlings. Bot. Gaz. 142(2):200-205.
13. Fitzgerald, C. H. and J. C. Fortson. 1979. Herbaceous weed control with hexazinone in loblolly pine (Pinus taeda L.) plantations. Weed Sci. 27(6):583-588.
14. Fisher, R. F. 1976. Allelopathic interference among plants: I. Ecological significance. Proc. Fourth N. Amer. For. Biol. Workshop. pp. 73-92.

256

15. Goggans, J. F. 1961. The interplay of environment and heredity as factors controlling wood properties in conifers with special emphasis on their effects on specific gravity. N. C. State Univ., Sch. For., Tech. Rpt. No. 11.

16. Hatzistathis, A. 1973. Soil moisture effects on stomata aperture, plant water stress and growth of Scotch and loblolly pine seedlings. Unpublished MS thesis. University of Georgia, Athens, GA.

17. Holt, H. A., J. E. Voeller, and J. F. Young. 1975. Herbaceous vegetation control as a forest management practice. Proc. South. Weed Sci. 28:219.

18. Horsley, S. B. 1976. Allelopathic interference among plants: II. Physiological modes of action. Proc. Fourth N. Amer. For. Biol. Workshop. pp. 93-136.

19. Kaufmann, M. R. 1968. Water relations of pine seedlings in relation to root and shoot growth. Plant Phys. 43:281-288.

20. Knowe, S. A., L. R. Nelson, D. H. Gjerstad, B. R. Zutter, G. R. Glover, P. J. Minogue, and J. H. Dukes, Jr. 1984. Fourth-year growth and development of planted loblolly pine on sites with competition control. South. J. Appl. For. (In press).

21. Larson, M. M. and F. W. Whitmore. 1970. Moisture stress affects root regeneration and early growth of red oak seedlings. For. Sci. 16:495-498.

22. McGregor, W. H. and P. J. Kramer. 1963. Seasonal trends in rates of photosynthesis and respiration of loblolly pine and white pine seedlings. Am. J. Bot. 50:760-765.

23. Michael, J. L. 1980. Long-term impact of aerial application of 2,4,5-T to longleaf pine (Pinus palustris). Weed Sci. 28(3):255-257.

24. Miller, A. E. and R. M. Allen. 1971. Effect of water stress on oxygen production of detached loblolly and white pine needles. Clemson Univ., College of For. and Rec. Res. For. Res. Series No. 22.

25. Nambiar, E. K. S. and P. G. Zed. 1980. Influence of weeds on the water potential, nutrient content and growth of young radiata pine. Aust. For. Res. 10:279-88.

26. Nelson, L. R., R. C. Pedersen, L. L. Autry, S. Dudley, and J. D. Walstad. 1981. Impacts of herbaceous weeds in young loblolly pine plantations. South. J. Appl. For. 5(3):153-158.

27. Pessin, L. J. 1939. Density of stocking and character of ground cover a factor in longleaf reproduction. J. For. 37:255-258.

28. _____. 1944. Stimulating the early height growth of longleaf pine seedlings. J. For. 42:95-98.

29. Preest, D. S. 1977. Long-term growth response of Douglas-fir to weed control. N. Z. J. of For. Sci. 7(3):329-32.

30. Robertson, C. F. and M. Reines. 1965. The efficiency of photosynthesis and respiration in loblolly pine. Proc. South. Conf. For. Tree Imp. 8:104-105.

31. Schmidtling, R. C. 1973. Intensive culture increases growth without affecting wood quality of young southern pines. Can. J. For. Res. 3:565-573.

32. Spurr, S. H. and W. Y. Hsiung. 1954. Growth rate and specific gravity in conifers. J. For. 52(3):191-200.

33. Spurr, S. H. and M. J. Hyvarinen. 1954. Wood fiber length as related to position in the tree and growth. Bot. Rev. 20:561-575.

34. Stransky, J. J. 1963. Needle moisture as mortality index for southern pine seedlings. Bot. Gaz. 124:178-179.

35. Stransky, J. J. and D. R. Wilson. 1964. Terminal elongation of loblolly and shortleaf pine seedlings under soil moisture stress. Soil Sci. Soc. Proc. 28:439-440.

36. Tappeiner, J. C., Jr. and S. R. Radosevich. 1982. Effect of bearmat (Chamaebatia foliolosa) on soil moisture and ponderosa pine (Pinus ponderosa) growth. Weed Sci. 30:98-101.

37. Tolley, L. C. and B. R. Strain. 1982. Effects of CO_2 enrichment and water stress on gas exchange of Pinus taeda (loblolly pine) seedlings. Plant Phys. 69(4):41.

38. Ursic, S. J. 1961. Tolerance of loblolly pine seedlings to soil moisture stress. Ecology. 42(4):823-825.

39. Walker, C. M. 1973. Rehabilitation of forest land. J. For. 71(3):136-137.

40. Walstad, J. D. 1976. Weed control for better southern pine management. Weyerhaeuser For. Paper No. 15. Weyerhaeuser Co., Hot Springs, AR. 44 p.

41. Zahner, R. 1962. Terminal growth and wood formation by juvenile loblolly pine under two soil moisture regimes. For. Sci. 8(4):345-351.

42. Zedaker, S. M. 1981. Growth and development of young Douglas-fir in relation to intra- and inter-specific competition. Unpublished PhD thesis. Oregon State University, Corvallis, OR. 175 p.

43. Zutter, B. R. 1983. Third year results-1980 pine herbaceous weed control growth impact. Auburn Univ. Silv. Herb. Coop. Res. Note No. 83-2. 6 p.

12. INTERFERENCE BETWEEN GREENLEAF MANZANITA (<u>ARCTOSTAPHYLOS PATULA</u>) AND PONDEROSA PINE (<u>PINUS PONDEROSA</u>)

S. R. RADOSEVICH

Associate Professor, Oregon State University, Corvallis 97331

ABSTRACT

Shrubs which usually occur on forested sites following a serious disturbance are believed to compete with newly planted conifers, thus restricting tree growth. When studying competition, it is important to recognize the importance of total plant density, proportions between species, and their spatial relationships. Several experimental designs (additive, substitutive, and systematic) have been used to study competition. In 1980 a substitutive experiment was established to examine the interaction between seedling ponderosa pine and sprouting greenleaf manzanita. This design considers all three factors since spatial arrangement and total density are held constant and proportion between species is varied. After the first growing season shrubs dramatically inhibited tree growth. Soil moisture was found to be the limiting factor for maximum tree growth when ponderosa pine occurred in association with greenleaf manzanita.

12. INTRODUCTION

Shrub-dominated brushfields often follow a serious disturbance to forest sites in the Sierra Nevada Mountains of California and many other regions of the western United States. The brushfields often are composed of several shrub species which represent an intermediate component of succession (3, 5). In Figure 1 several stages of shrub occupation following clear-cut logging are depicted. The site usually is invaded quickly by seedling or sprouting shrub species that are believed to act as competitors to newly planted conifers.

Duryea, M.L. and Brown, G.N. (eds.). Seedling physiology and reforestation succes
©1984, Martinus Nijhoff/Dr W. Junk Publishers, Dordrecht/Boston/London. ISBN 978-94-009-6139-5

260

FIGURE 1. Secondary succession after logging activity.
Various stages of community development are evident depending
on the time (years) after conifer canopy removal (8).

Stewart (9) has summarized over 100 studies which report
the coniferous responses to brush control. Such studies
generally are instructive because they demonstrate the relative aggressiveness of brush species in competition with
conifer seedlings. However, it is difficult to determine
from those studies the levels of shrub suppression necessary
for optimum conifer growth, or to assess times in stand
development when competition is most critical. In addition,
few studies have attempted to identify the environmental
resource(s) for which competition occurred. The objective of
this paper is to examine the important design and environmental parameters that are necessary to consider when conducting
research on competition. In addition this paper describes
the mutual responses of greenleaf manzanita and ponderosa
pine to interspecific interference and to the resource
availability that resulted from shrub manipulation.

12.2 METHODS TO STUDY COMPETITION
The study of competition is complex because several
interactive features of the plants and their environment must
be considered. Shrub and tree growth are influenced by the
abundance of various environmental resources. Furthermore,
the relative ability of each species to capture or usurp a
resource influences the availability of that resource for the

growth of associated species. Plant species also can respond
to resources in different ways. Thus a knowledge of environmental resources and plant physiological responses to them
are necessary to understand most competitive interactions.

The environmental resources that are usually most significant in competitive interactions are light, water, and/or
soil nutrients since each is consumed directly by plants for
growth. The gases (CO_2 and O_2) which are involved in photosynthesis and respiration usually are not competed for
directly by plants. However, these gasses are substrates for
important physiological processes and may be indirectly
involved in competition since both photosynthesis and
respiration are influenced by the abundance of other resources (e.g., water and light). Temperature is not considered
to be an environmental resource, because it is not consumed
by plants for growth. However, temperature does influence
plant productivity and can be a significant factor in competitive interactions. For example, Conard and Radosevich
(1) observed that maximum air temperatures wre consistently 3
to 4°C higher within a Ceanothus velutinus canopy than above
it. Based on temperature studies of CO_2 fixation with white
fir (Abies concolor) and C. velutinus, Conard and Radosevich
(1) found that such an increase in air temperture could
reduce white fir productivity and the outcome of competition
between the two species.

In order to adequately study competition, it is necessary
to consider several other factors in addition to environmental resources and conditions. These factors are: total
plant density, the proportion of each species in the mixture,
and the spatial arrangement of each species relative to the
other. There are several experimental designs that have been
employed to study competitive interactions which recognize
the above factors to various degrees. The most common type
of experimental design for competition study is the additive
experiment. In additive experiments two species are grown
together, with the density of one species held constant while
the density of the other species is varied. Thus, one

species acts as the biological indicator of the aggressive
ability of the other. The approach is used widely because of
its apparent relevance to field situations, where one species
is established in an area at a fixed density (e.g., a forest
plantation) and the area then is invaded by another. This
approach can provide insight into the costs of competition in
terms of yield loss. However, it is difficult to determine
the degree of interaction either between species or among
individuals of the same species. Furthermore, it is
impossible using this techinque to actually determine which
species is most competitive. The value of the additive
approach is the ability to determine directly, the costs
(yield loss, etc.) that are associated with the absence of
weed control (4, 8).

Other approaches have been developed which more adequately consider the experimental factors of total density,
proportionality, and spatial arrangement than additive studies. These approaches have been termed substitute and
systematic designs. Neither approach has been used extensively to study competitive interactions in forest systems.
The substitutive approach relies heavily on the concept that
total plant yield is independent of plant density; i.e., that
each site has a finite carrying capacity for vegetation. A
basic assumption of the substitutive approach is that the
yield in mixture can be determined from the yield of each
species when grown separately (4, 2). The substitutive
approach (replacement series) requires that the total density
or final yield be constant and that the two species be grown
at varying proportions ranging from 0 to 1.0 (2). Therefore,
the deviation in the actual from expected yield of each species when grown in the various proportions defines the nature
of any interspecific interaction. Harper (4) described four
basic types of interactions which may occur from plant to
plant association (Figure 2). These are no interference (a),
competition (b), mutual antagonism (c), or mutual benefit
(d). These possible interactions have been discussed in
greater detail in other texts (4, 8) than is possible to

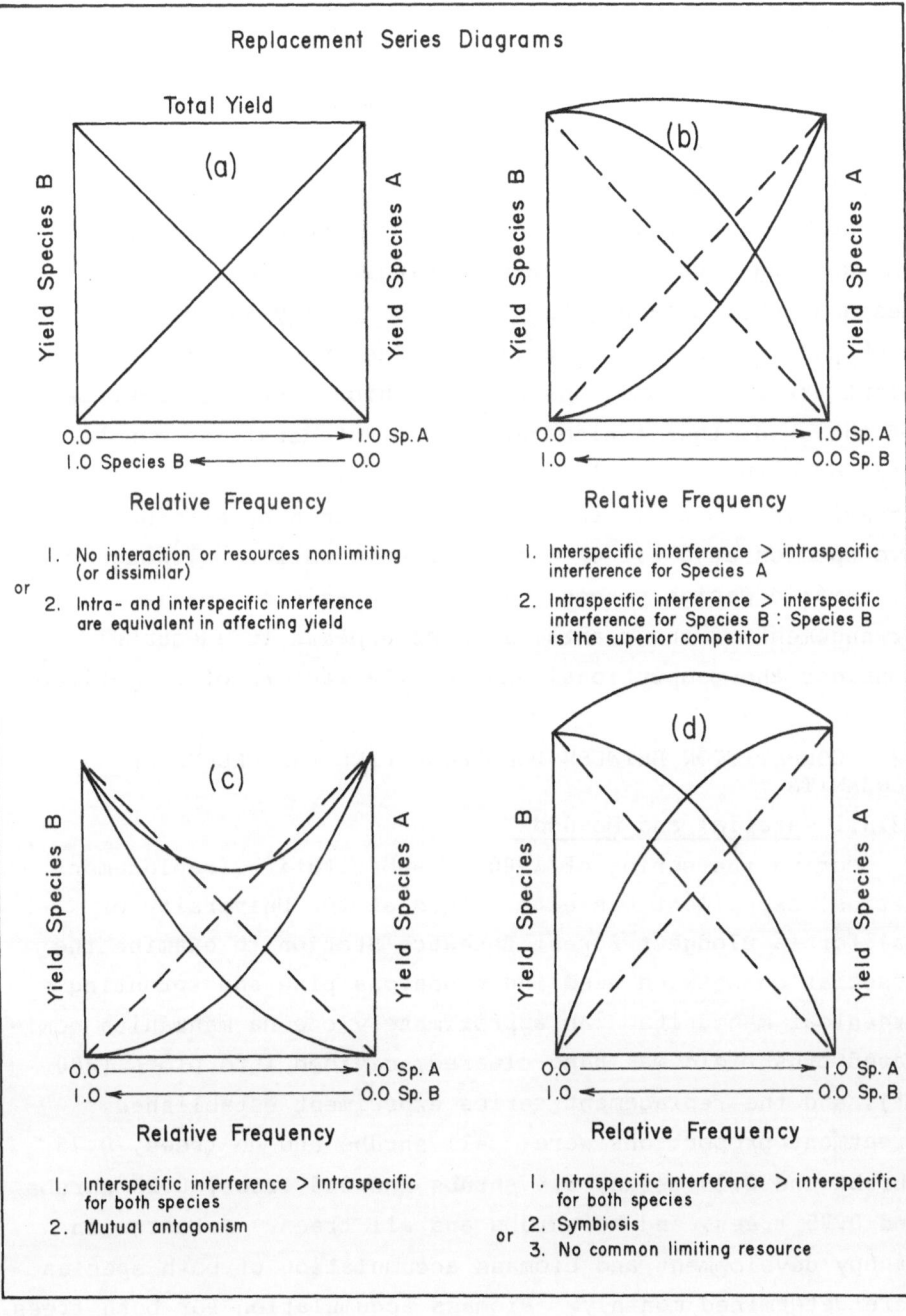

FIGURE 2. A variety of models for results of replacement series experiments for interference study. The vertical axis indicates some measure of plant yield, while the horizontal axis represents the proportion (0 to 1.0) of the two species in mixture (4).

accomplish here. Although the substitutive approach is
seemingly artificial to field situations where crop (tree)
densities are fixed, it appears to provide a more accurate
assessment of competitiveness than can be achieved using the
additive approach.

Systematic designs were first considered by Nelder (7)
for spacing studies involving a single species. These
designs consist of a grid of points usually as an arc, with
each point representing a plant. The space available to each
plant varies in some consistent fashion over different parts
of the grid, thus a wide range of plant densities can be
studied without changing the pattern of arrangement.
Interference between species is introduced by alternating the
two species at a 1:1 or other suitable ratio. Although the
systematic design places greatest emphasis on spatial
arrangement of the plants, it also appears to adequately
consider the proportional and density factors of competition.

12.3 COMPETITION BETWEEN PONDEROSA PINE AND GREENLEAF MANZANITA

12.3.1 Material and Methods

During the spring of 1980, a substitutive (replacement
series) experiment was established at the University of
California Blodgett Forest Research Station to examine the
association between seedling ponderosa pine and sprouting
greenleaf manzanita. An approximately one ha manzanita dominated brushfield was hand-cleared, divided into plots (300
m^2), and the replacement series experiment established.
Treatment proportions were: all shrubs and no trees, 0.75
shrubs and 0.25 trees, 0.5 shrubs and 0.5 trees, 0.25 shrubs
and 0.75 trees, and no shrubs and all trees. Patterns in
canopy development and biomass accumulation of both species
were determined monthly. Biomass accumulation for both trees
and shrubs was estimated from calculated values of canopy
volume by dimensions analysis (6). Fluctuations in soil
moisture (0.3 to 1.5 m in depth), using a neutron probe, were
measured monthly throughout each growing season.

12.3.2 Results and Discussion

High correlation (r = 0.97 and 0.99 for greenleaf manzanita and ponderosa pine, respectively) between canopy
volume and above-ground biomass were observed for both plant
species (Figure 3). Therefore, data in the following
discussion will be presented on a dry weight basis, even
though the plants were not destructively harvested during the
experiment.

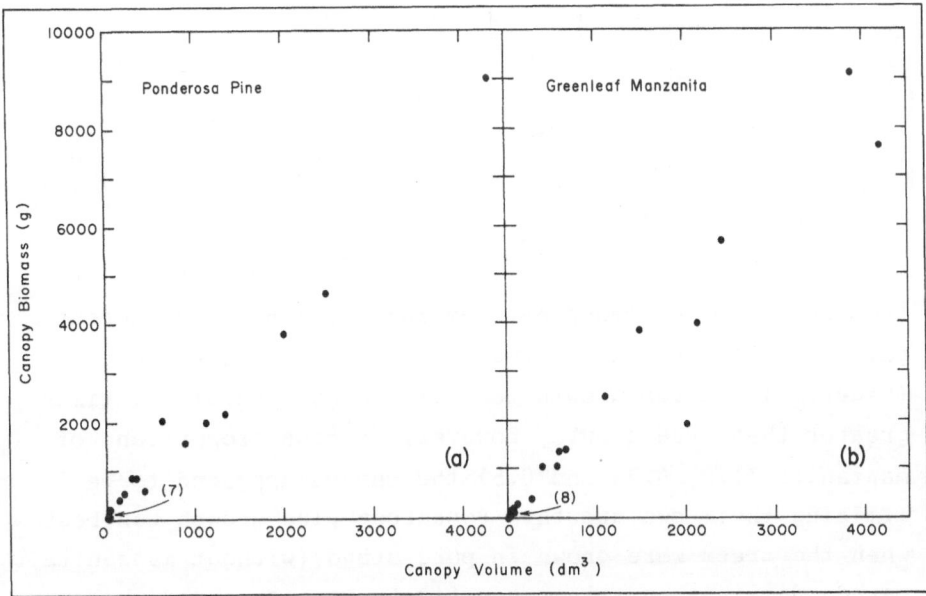

FIGURE 3. Regression of canopy volume to above-ground
biomass for ponderosa pine (A) and greenleaf manzanita (B).
Canopy volume was derived for ponderosa pine from measurements of tree height and canopy diameter, using the equation
for a right cone. Canopy volume for manzanita was derived
from measurements of shrub diameter using the equation of a
sphere (10). Regression equations for ponderosa pine and
greenleaf manzanita are y = 1.97x + 22.8 and y = 1.99x +
20.5, respectively.

During the first growing season (1980), no interference
was observed between shrubs and trees (Figure 4). However,
approximately 60 times more shrub growth than tree growth
occurred during that time. The lack of interference observed
during the first growing season was probably a result of the
relatively large distance which existed between individual

266

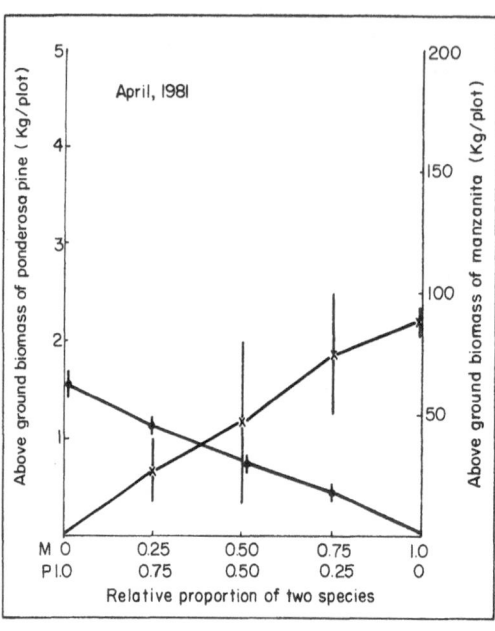

FIGURE 4. Biomass accumulation in April 1981, of ponderosa pine (●) and greenleaf manzanita (x) when grown in replacement series.

tree and/or shrub canopies. By fall of 1982, a different pattern of interaction between the species was evident (Figure 5). Shrub growth was still approximately 35 times greater than tree growth, however, at high proportions of manzanita (1.0, 0.75 and 0.5) the shrubs appeared to be limiting their own growth. Ponderosa pine growth was best when the trees were grown in pure stand (without manzanita). Marked declines in tree productivity were observed whenever shrubs were grown in association with trees. A manzanita proportion of only 0.25, that of the pure shrub stand, resulted in nearly a 60% loss in tree productivity (Figure 5). Higher levels of shrub occupancy resulted in 80% to 90% reductions in tree growth. These observations have clear implications to forest managers who are seeking to optimize conifer productivity in newly planted forests. It appears that very large reductions in shrub biomass are necessary to maintain productive young plantations (Figure 5).

An important index of plant productivity is the relative growth rate (RGR). It is an overall growth index for comparing the rate of plant growth at different times, among different populations, or among different environments

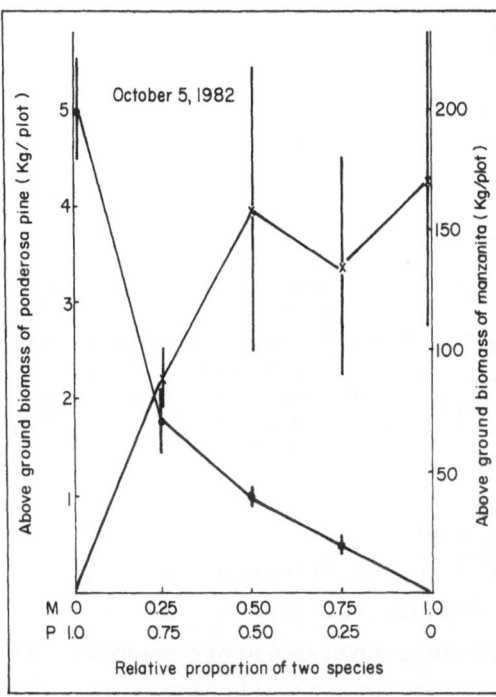

FIGURE 5. Biomass accumulation on October 5, 1982, of ponderosa pine (●) and greenleaf manzanita (x) when grown in replacement series.

(treatments). Because absolute growth depends on the amount of growing material, Ledig (6) indicates that it would be meaningless to compare growth rates among individuals that differ in size. Therefore, growth rate is expressed relative to the amount of growing material already present. RGR, the change in dry weight (W) per unit of time (t) per unit of growing material is given by the following equation:

$$RGR = (\frac{\Delta W}{\Delta t})(\frac{1}{W})$$

The RGR values of ponderosa pine growing in pure stand and at various proportions of shrubs are presented in Figure 6 for the 1982 growing season. No differences in relative growth rates between treatments were observed for the entire growing season when trees and shrubs were grown together. This is in contrast to the marked increase in RGR of ponderosa pine without shrub interference (Figure 6). Clearly, growth of ponderosa pine is inhibited by association with the shrub species and competition is most critical during the spring and summer portions of the growing season.

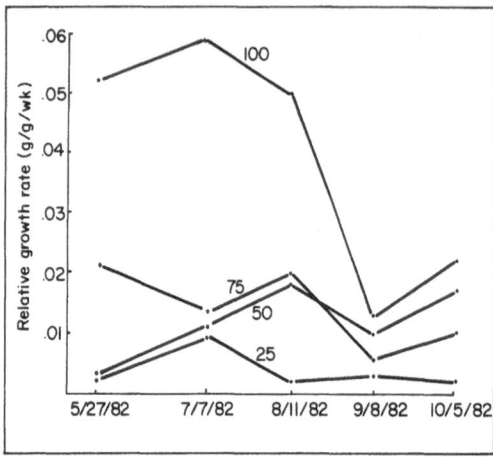

FIGURE 6. Relative growth rates (RGR) during 1982 of seedling ponderosa pine growing in pure stand and associated with green-leaf manzanita. 100, 75, 50 and 25 refer to relative proportions of: all trees, 0.75 trees and 0.25 shrubs, 0.50 trees and shrubs, and 0.25 trees and 0.75 shrubs, respectively.

The soil moisture potential of each treatment (proportion) is shown at the 0.3 m and 1.5 m depths (Figures 7 and 8, respectively). At both soil depths, the presence of shrubs dramatically depleted the available soil moisture supply. In 1982, the single treatment that was grown without shrubs, i.e., all trees, maintained soil moisture (1.5 m) near field capacity throughout the growing season. This was in spite of the much higher tree density established in that treatment. The shrubs present in this experiment made a very large and significant demand upon the water resource of the site. When all environmental data were subjected to multiple regression analysis, soil moisture availability at the 1.5 m level was the most critical factor determining conifer growth.

These data, when considered in total, indicate that ponderosa pine growth is severely limited by the presence of sprouting greenleaf manzanita. Furthermore, the decline in tree growth is caused by rapid use of water by the shrubs. In order to optimize tree productivity in new plantations, forest managers must manipulate shrub stands in a manner that will make available to trees a significant amount of the water resource. Thus, in the Sierra Nevada, shrub stands must be manipulated in a severe enough manner to make available a significant portion of the water resource to the trees. It also is important to realize that information of this type is necessary for foresters to effectively predict

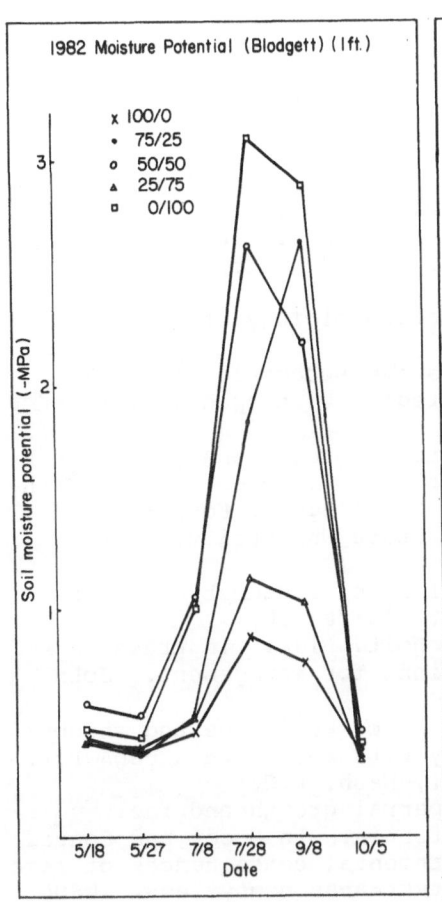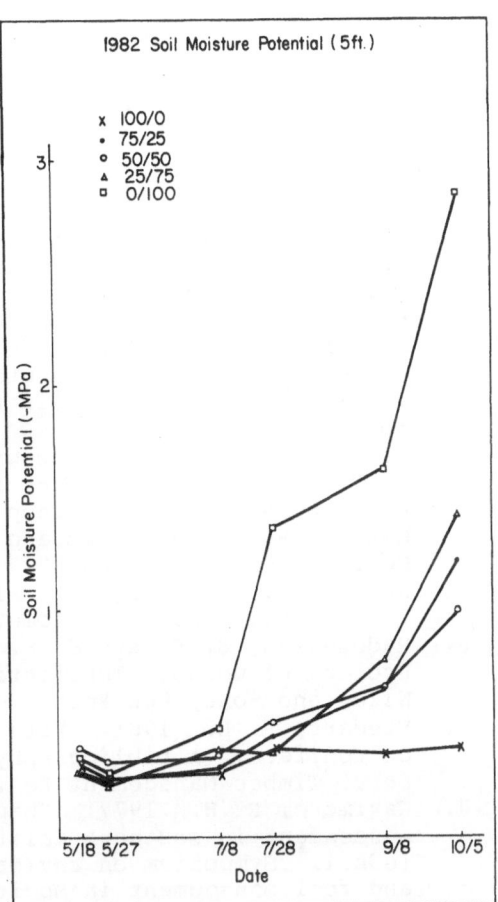

FIGURE 7. Soil moisture potential (0.3 m) during 1982. Greenleaf manzanita and ponderosa pine were grown in replacement series. x, ●, o, ∧, refer to relative proportions of: all tree, 0.75 shrub and 0.25 tree, 0.50 shrub and tree, 0.25 shrub ans 0.75 tree and all shrub treatment, respectively.

FIGURE 8. Soil moisture potential (1.5 m) during 1982. Greenleaf manzanita and ponderosa pine were grown in replacement series. x, ●, o, ∧, refer to relative proportions of: all tree, 0.75 shrub and 0.25 tree, 0.50 shrub and tree, 0.25 shrub and 0.75 tree and all shrub treatments, respectively.

the consequences of shrub occupancy on conifer yields or to determine the appropriate degree of vegetation manipulation (resource availability) required for optimum tree growth.

REFERENCES

1. Conard, S. G. and S. R. Radosevich. 1982. Growth responses of white fir to decreased shading and root competition by montane chaparral shrubs. For. Sci. 28: 309-320.
2. deWitt, C. T. 1960. On competition. Verslagen van landbouwkundige onderzorigen. No. 66.8.
3. Dyrness, C. T. 1973. Early stages of plant succession following logging and burning in the western Cascades of Oregon. Ecology 54:57-69.
4. Harper, J. L. 1977. Population biology of plants. Academic Press, London.
5. Issacs, L. A. 1940. Vegetative succession following logging in the Douglas-fir region with special reference to fire.
6. Ledig, F. T. 1974. Concepts of growth analysis. In C. P. P. Reid and G. H. Fechner (eds.). Proc. third North Amer. forest biol. workshop. College of Forestry and Natural Resources, Colorado State University, Fort Collins, CO. pp. 166-182.
7. Nelder, J. A. 1962. New kinds of systematic designs for spacing studies. Biomet. 18:283-307.
8. Radosevich, S. R. and J. S. Holt. 1984, in press. Ecology of weeds: Implications for management. John Wiley and Sons, New York.
9. Stewart, R. E. 1981. Effects of weed tree and shrubs on conifers: A bibliography with abstracts. USDA For. Serv. Timber Management Res., Wash. D.C.
10. Wakimoto, R. H. 1977. Chaparral growth and fuel assessment in southern California. In Moody and Conrad (eds.). Symposium on enviornmental consequences of fire and fuel management in Mediterranean ecosystems. USDA For. Serv. Gen. Tech. Report No. 3. pp. 412-419.

13. NUTRITION MANAGEMENT: A PHYSIOLOGICAL BASIS FOR YIELD IMPROVEMENT

J. T. FISHER AND J. G. MEXAL

Assoc. Professor and Head, Department of Horticulture, New Mexico State University, Box 3530, Las Cruces, New Mexico 88003

ABSTRACT

As resources become more limiting, greater emphasis will be placed on nursery and forest nutritional management to achieve reliable reforestation, shorten rotations and improve marginal site productivity. Nursery fertilization with few adverse effects can increase production efficiency. Properly applied fertilizers increase seedling growth after outplanting and enhance seedling drought and cold-hardiness.

Forest fertilization during early stand age can exploit favorable site, genotype, and operational interactions. Preplant fertilization frequently increases growth over the entire rotation. In the South, delayed fertilization often results in slow growth and seedling mortality. Large gains in early forest growth and survival will require successful vegetation control and the integration of the numerous biological and physical factors governing fertilization response.

13.1 INTRODUCTION

The demand for wood and wood products is projected to increase well into the next century, primarily the result of population growth. A major problem in meeting this demand is the lack of additional productive lands. To avoid serious wood shortages in the future, foresters must move from extensive to intensive forest management on existing commercial forest land, and successfully reclaim low quality sites. Their management objectives will be to optimize yield per unit area of land. However, as foresters try to increase wood fiber production,

Duryea, M.L. and Brown, G.N. (eds.). Seedling physiology and reforestation succes
©1984, Martinus Nijhoff/Dr W. Junk Publishers, Dordrecht/Boston/London. ISBN 978-94-009-6139-5

the resources necessary for optimum growth and utilization will become more limiting.

Forest soils can be drained of essential plant nutrients under intensive management. Fertilization is, therefore, needed to restore and maintain soil nutrients above critical levels. Fertilization in the nursery can impact yield throughout the life of the plantation. Proper fertilization increases transplant performance and fertilization of immature stands results in substantial growth responses. In older stands, fertilization can balance height class distributions and increase volume growth. In addition, strategic fertilizer use can increase the potential gain from tree improvement programs, soil management practices and thinning regimes.

Fertilization costs and benefits are important issues that will demand closer attention as worldwide competition and demand for fertilizers increase. Despite a slight decline (1%) in use in 1981, fertilizer consumption will increase and the United States will receive a smaller share of the future world supply. Our understanding of nursery and forest nutrition must, therefore, advance to produce the most growth at the lowest possible cost.

The biological, chemical and physical factors impacting fertilization responses are complex and highly interactive (Fig. 1). This paper reviews the role of nursery and forest fertilization in seedling establishment and early growth.

13.2 NURSERY SEEDLING NUTRITION

Response to nursery fertilization is influenced by a number of factors including soil and plant nutritional status, bed density, length of growing season, soil organic matter, previous cropping practices and seedling genotype. Several reviews of the nutritional status of tree seedlings and nursery soils provide excellent discussions of important factors receiving minor or no attention here (3, 57, 85, 94). These include deficiency symptoms, mineral contents, and soil and plant chemical testing. Only factors directly related to yield improvement will be discussed.

273

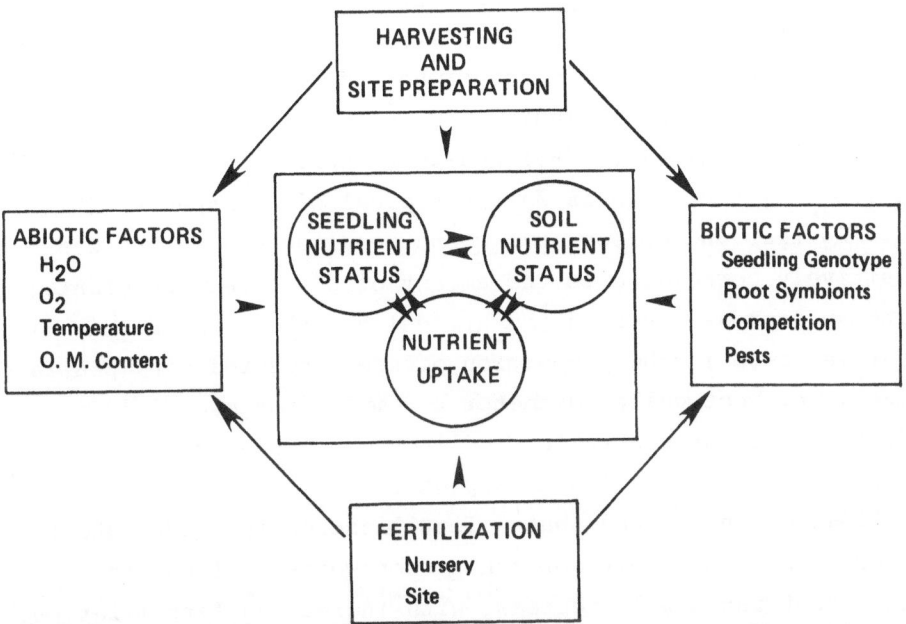

FIGURE 1. Model of factors influencing forest stand nutrient status.

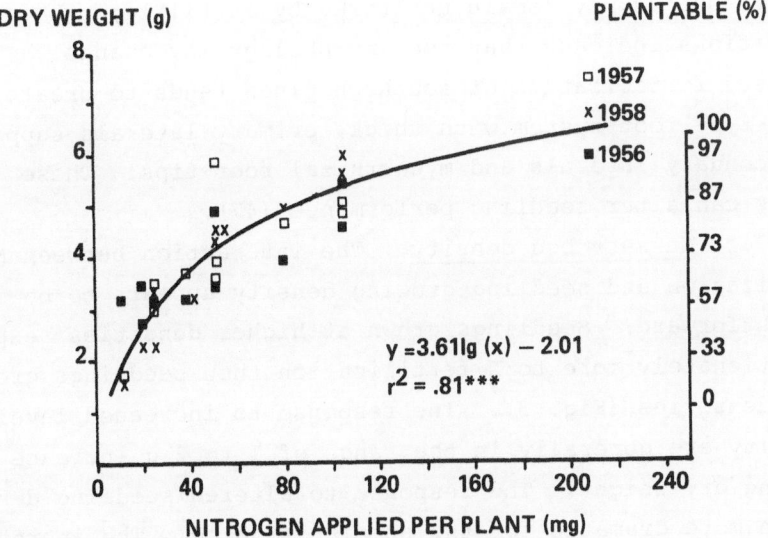

FIGURE 2. Effect of nitrogen fertilization on loblolly pine seedling dry weight (adapted from 87).

13.2.1 <u>Seedling biomass and distribution</u>. Nitrogen (N) is
commonly applied to nursery soils to increase seedling size.
The response of several major timber species to N fertilization
has been evaluated, including lodgepole pine (93), loblolly
pine (87), Douglas-fir (95) and white spruce (3). The response
of loblolly pine to N is fairly typical of all species (Fig.
2). Switzer and Nelson (87) found seedling dry weight was
positively correlated to the amount of N applied per plant.
With an increase in seedling dry weight, there was a concomitant increase in the proportion of seedlings reaching plantable
size (i.e. root collar diameter $\geq$ 3 mm). Maximum yield in
terms of percent plantable seedlings was achieved with an
application of about 160 mg N/plant.

It is often assumed that N fertilization increases shoot
weight proportionately more than root weight. Thus, the
root:shoot ratio will decrease with increasing fertilization.
This imbalance could adversely affect field performance.
However, data on loblolly pine (87), Douglas-fir, Sitka spruce
and lodgepole pine (95) indicate fertilization has little, or
at worst, minimal negative impact on root:shoot ratio. Van den
Driessche quadrupled nitrogen fertilization to show a significant decrease in root:shoot ratio. While the absolute ratio of
roots to shoots may remain unaltered by fertilization, casual
observations indicate that root morphology may change.
Increased fertilization of southern pines tends to create a
carrot-type root system with thick, primary laterals supporting
few secondary laterals and mycorrhizal root tips. These
factors can alter seedling performance (47).

13.2.2 <u>Nursery bed density</u>. The interaction between N
fertilization and seedling growing density appears to be
straightforward. Seedlings grown at higher densities respond
proportionately more to N fertilization than seedlings grown at
lower densities (Fig. 3). The response to increased levels of
fertility are generally in the range of 1 to 2 g increase in
seedling dry weight. The response to altered seedling density
is much more dramatic for the species studied. The increase in
seedling biomass is of the order of 5 g, over the range of seed
bed densities studied.

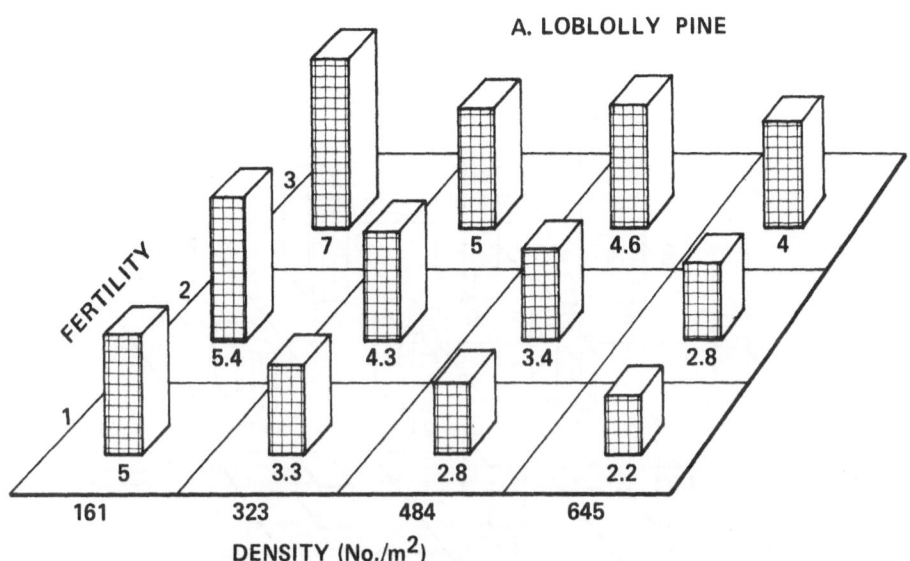

FIGURE 3 a. Loblolly pine seedling dry weight (g) response to growing density (No./m²) and fertility (adapted from 87).

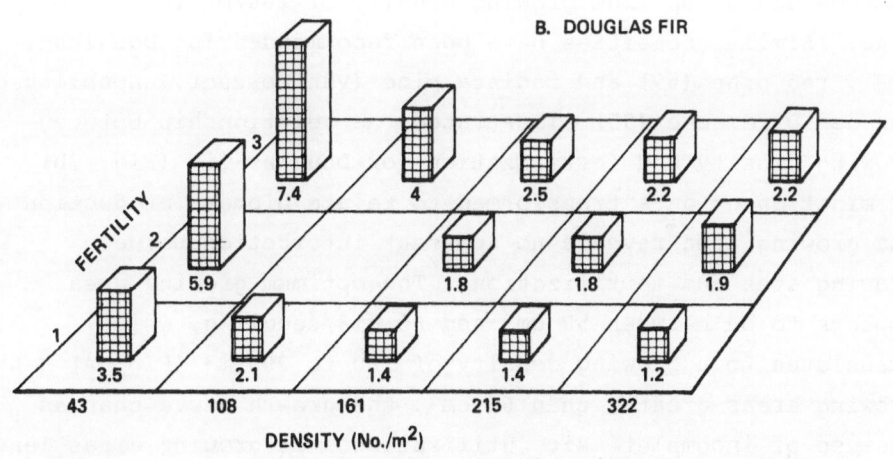

FIGURE 3 b. Douglas-fir seedling dry weight (g) response to growing density (No./m²) and fertility (adapted from 95).

276

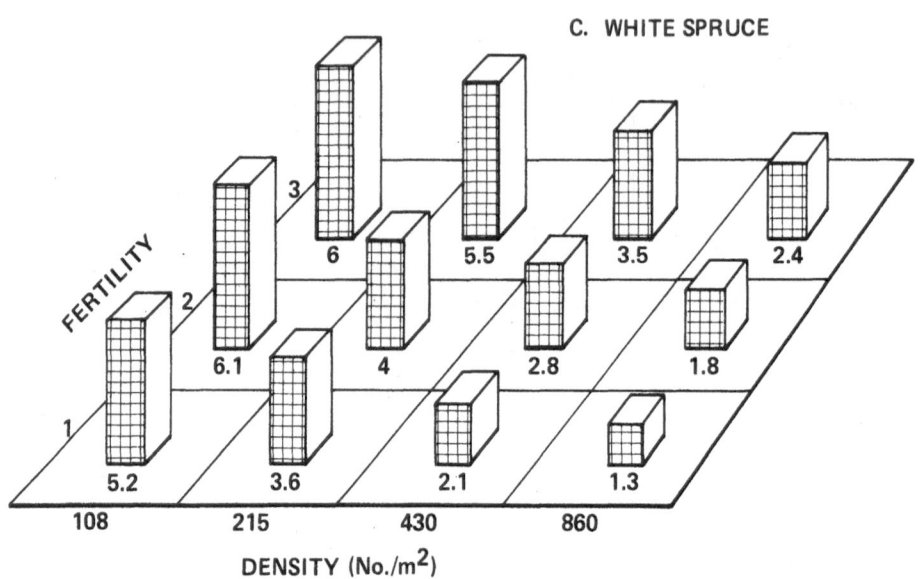

FIGURE 3 c. White spruce seedling dry weight (g) response to growing density (No./m²) and fertility (adapted from 3).

The optimum growing density does not appear to be influenced by the fertility level. The optimum growing density offers a trade-off between maximizing seedling size and the number of outplants produced per unit area of nursery bed. Mexal (55) recommended an optimum growing density of 200/m² for loblolly pine. Similar densities have been recommended for Douglas-fir (28), red pine (59) and radiata pine (Van Dorsser, unpublished). Van den Driessche (95) illustrated the relationship between growing density and fertilization for Douglas-fir (Fig. 3b). Examination of data transformed to relate biomass production and growing area reveals no apparent interaction between growing area and fertilization. The optimum growing area appears to be between 50 cm² and 60 cm²/seedling, which translates to a growing density of 180 to 200/m² (Fig. 4). At growing areas greater than 60 cm², the growth curve changes because of incomplete site utilization. At growing areas less than 50 cm², competition increases and individual seedling weight decreases.

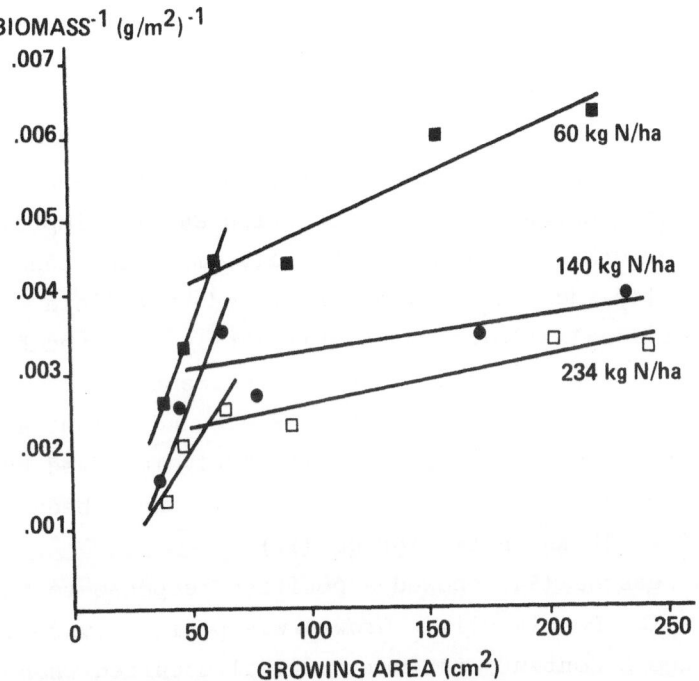

FIGURE 4. Interactive effects of growing area and fertility on biomass production (g m^{-2})$^{-1}$ (adapted from 95).

13.2.3 <u>Cold hardiness and fall fertilization</u>. Once seedlings have achieved target size, fertilization is usually stopped and irrigation decreased to induce dormancy. Potassium chloride (KCL) may be applied to promote cold-hardiness; however, there is ample data to indicate potassium (K) fertilization has little, if any, effect on cold injury (62). More importantly, a balanced nutritional regime is required for maximum cold-hardiness (90). While K is not required for cold-hardiness, it is likely that K is required for outplanting success through regulation of plant water status (16, 22). Potassium regulates stomatal control and drought tolerance of transplanted seedlings. Low K levels in the foliage may result in loss of stomatal control during the critical establishment period. Potassium is highly mobile in the soil and deficiency levels can develop in sandy soils at season end. This would warrant application of additional K in the fall. Seedling fertilization with KCL in the fall should continue, but for entirely different reasons.

Fall fertilization with N has received minor attention, especially in recent years. Woody plants, especially conifers, continue biomass accretion during the dormant season (63, 92). Root growth of many conifers continues as long as soil temperature is above 5° C. Fertilization with low levels of N after budset can increase biomass accretion and may improve outplanting performance, without adversely impacting seedling height. Hinesley and Maki (40) demonstrated that field performance of longleaf pine was significantly improved by fall fertilization in the nursery. Survival, percentage of longleaf pine seedlings emerging from the grass stage and height at 8 years following outplanting were significantly improved by fertilization before lifting. Similar results have been shown for Douglas-fir (2) and Sitka spruce (12).

Van den Driessche (94) showed a positive response to N fertilization for Douglas-fir. Growth was positively correlated with foliage N content; maximum survival occurred when foliage contained 2% N. He recommended applications of 50 to 75 kg N/ha the first year, and 100 to 150 kg N/ha the second year. Fall fertilization was not discussed, but may provide an opportunity of regulating N content without contributing to seedling gigantism (87).

13.2.4 <u>Field Response to Nursery Fertilization.</u> Nursery cultural practices can have long term influences on forest yield due to the impact of initial seedling biomass on early performance (96). Field survival is not often correlated with the level of nursery fertility. However, field growth and nursery nutritional status are clearly related (87, 95). Autry (4) reported a positive correlation between biomass at planting and individual tree volume at age 14 using plantations established from the studies of Switzer and Nelson (87) (Fig. 5). Biomass at time of planting accounted for 66% of the variation in tree volume at age 14. Volume was increased 38% across the seedling size range at the time of planting. It is reasonable to assume similar responses can be expected for all the major timber species.

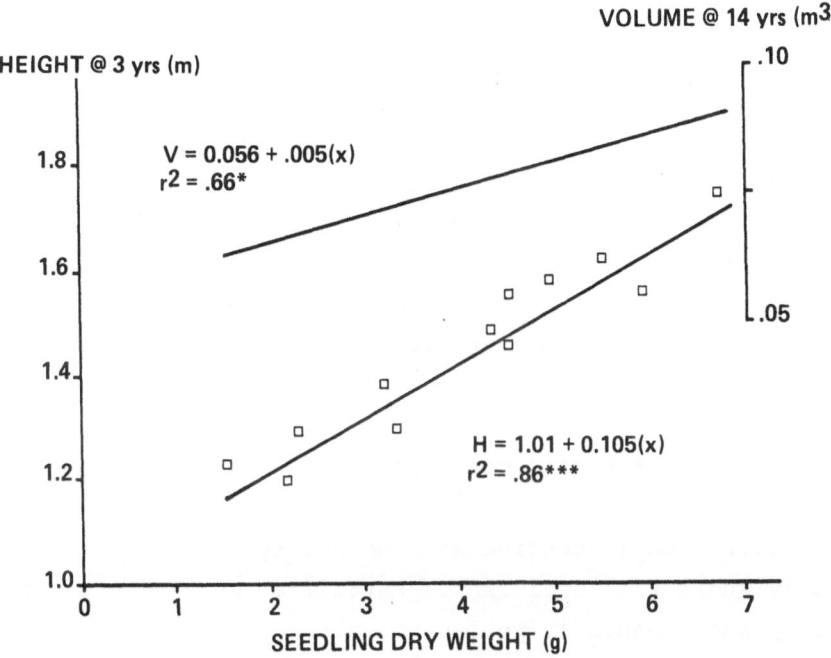

FIGURE 5. Relationship between initial seedling dry weight and height after 3 years and individual tree volume after 14 years in the field (adapted from 4 and 87).

13.3 EARLY PLANTATION FERTILIZATION

Fertilization at time of planting or shortly thereafter may be required to alleviate soil nutrient deficiencies or to speed seedling growth where nutrient levels are suboptimal. Forest soils are often deficient in certain essential elements, especially N. Adequate reserves may be present in some soils, but in forms unsuitable or unavailable for plant growth. Nitrogen deficiencies are most often reported in coniferous forests of cold climates under conditions that favor accumulation of thick acid humus (99). Incipient N deficiencies are also found in many sandy soils of warmer climates, including the flatwoods and sandhills of the south-eastern coastal plain and the Douglas-fir region of the Pacific Northwest (67). Phosphorus (P) deficiencies occur in wetlands of the coastal plains of the United States and peat bogs of boreal forests (45, 68, 97, 100). Potassium often is limiting on sandy soils (44). In Great Britain, forests established on poorer soils, notably acid peat, require only P to produce reasonable early growth, but eventually K is required for vigor. Geologically old soils, such as those in Australia, may be limiting in a number of nutrients, especially micronutrients (74). Zinc, sulfur and boron are routinely applied when specific deficiencies require corrective measures.

The period following establishment is marked by rapid growth and maximum rate of nutrient uptake. Preplant fertilization may improve growth over the entire rotation. In many cases, survival of planted seedlings is significantly improved; in extreme cases, fertilization may be essential to successful stand establishment.

In the Pacific Northwest, most operational fertilization has taken place on younger stands with closed crowns. In the South, fertilization is considered a necessity on many sites, particularly where P is deficient. Fertilization often results in dramatic, and frequently permanent, alteration of site characteristics (Fig. 6). A delay in fertilization results in slow tree growth and often in poor survival. Small amounts of N, applied in combination with P, usually result in better growth than P alone. The importance of P nutrition on coastal plain soils was illustrated by Burns and Hebb (19) who found a 4 m difference in height between controls and trees fertilized 10 years previous. Results indicate P fertilization has a long lasting effect and permanently changes the relative growth rate of the stand (Fig. 7). Ballard (6) has shown that fertilization can impact growth response in second rotation plantations as well. Nitrogen is rarely applied to plantations prior to crown closure. However, seedling response to preplant N fertilization is frequently positive (Table 1). Species with monocyclic or determinate shoot growth patterns generally respond to N. Species with polycyclic or indeterminate growth do not respond without vegetation management. In fact, N is detrimental without weed control because of allelopathy or competition for soil water.

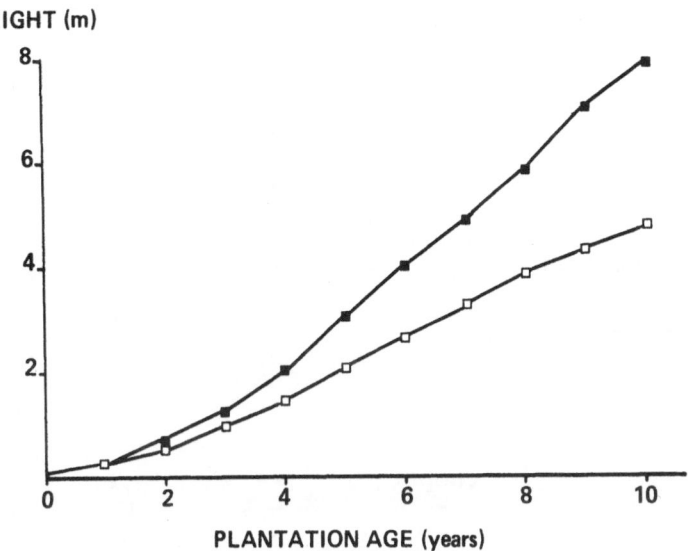

FIGURE 6. Height growth response of slash pine to preplant phosphorous fertilization (19).

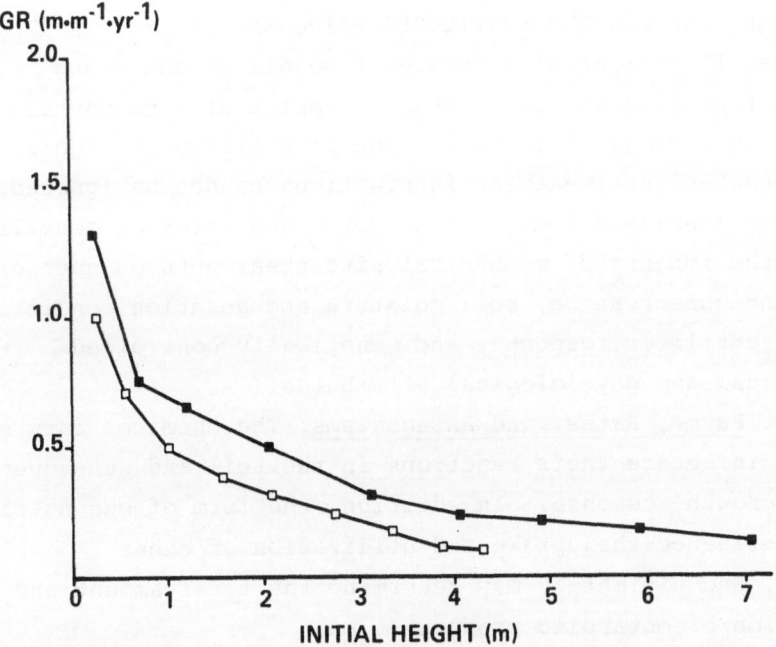

FIGURE 7. Relative growth rate (RGR) of slash pine in response to preplant phosphorous fertilization (drawn from Figure 6). Initial height refers to height at the beginning of each growing season.

Table 1. Response of conifers to preplant fertilization with N.

Shoot Growth Pattern	Site Prep	Response	Fertilizer	Citation
Monocyclic				
Picea abies	-	+	N,P,K	37
P. glauca	skalp	+	N,P,K	13
P. abies	-	+	N,P,K	31
P. sitchensis	-	+	N	26
Pinus contorta	skalp	+	N,P,K	13
Pseudotsuga menziesii	-	-	N	38
P. menziesii	-	+	OSMOCOTE	20
Tsuga heterophylla	-	+	OSMOCOTE	21
Polycyclic				
Pinus elliottii	bedding	+	N	5
P. elliotti	-	-	N	9
P. radiata	chop	+	N,P	7
P. radiata	-	-	N	29
P. taeda	-	-	N	9
P. taeda	-	+	N,P,K	66
P. taeda	-	-	N,P,Ca	58
P. taeda	irrigate	-	N,P,Mg	18

13.4 FACTORS INFLUENCING FERTILIZER RESPONSE

The overall biological response of seedlings and young stands to fertilization integrates a complex of site physical, chemical and biological factors. The probability of significant pre- and post-plant treatment interactions cannot be ignored. Among major considerations are the forms and rates of fertilizers applied, the impacts of mechanical site treatments on nutrient cycling and conservation, soil moisture and aeration conditions limiting fertilizer response, and genetically controlled morphological and physiological attributes.

13.4.1. Forms, Rates, and Antagonisms. The chemical form of nutrients influence their reactions in the soil and subsequent seedling growth response. In addition, the form of one nutrient can influence the uptake and utilization of other nutrients, and ultimately may determine the total amount and distribution of metabolic products.

Forms in which N is present and the kind of associated ions in the fertilizer compound can produce effects which may be more important than considerations of N concentration. For example, ammonium sulfate is superior to other forms of N for growth of young conifers (11, 52, 64). Ammonium sulfate has been more effective than ammonium nitrate in reforesting denuded areas, such as old borrow pits (67).

Growth response of conifers to a particular nutrient is governed to a great extent by the rate and method of application. Maintenance of stable nutrient levels is a major challenge on forest sites. Development of accurate and closely controlled nutrient application techniques, particularly for N, should receive greater attention. Because of the exponential relation between seedling weight and N status, a strict control over N fluctuations is necessary to maintain maximum growth (43). Typically, fertilization every 5 years might yield a response curve similar to that illustrated in Figure 8. However, it is conceivable the application of the same total amount of fertilizer at 3 year intervals could maintain the peak level of response, increasing wood production. This concept, as proposed by Ingestad (42), appears feasible on an operational scale. The burden of increased applications would be offset by increased land area that would receive fertilizer at each interval.

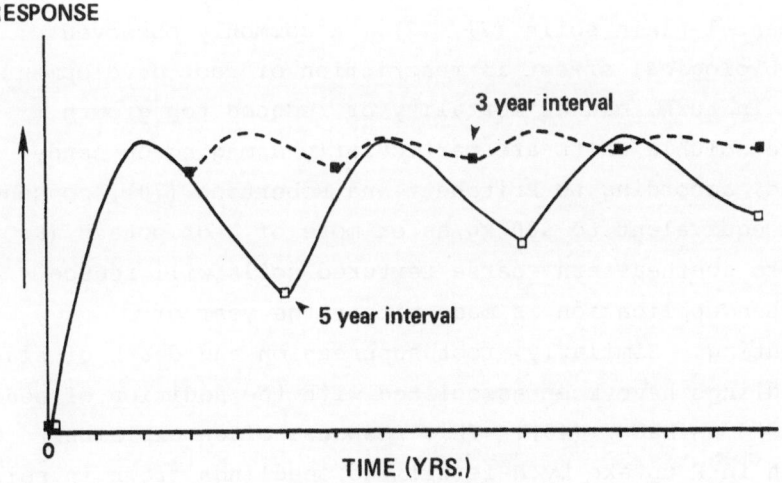

FIGURE 8. Idealized response to fertilization frequency (after Ingestad 42).

Unbalanced ratios of N, P and K lead to disturbance of nutritional conditions and assimilation processes, and eventually are reflected in deficiency symptoms and yield losses. This underscores the need to carefully monitor foliage nutrient levels following fertilization. Optimal N/P ratios for tissues and fertilizers have been reported (14, 53). Brix and van den Driessche (17) suggested nutrient proportions for good growth of north temperate zone conifers. Even small imbalances can depress growth. The effects of high N rates on decreasing the root:shoot ratio were eliminated by adequate levels of P fertilization (41). High levels of N restrict mycorrhiza development which, in turn, results in poor P absorption (75). Nitrogen influences P availability through numerous biological and chemical effects (34). Biological effects are those caused indirectly by N on the form and function of the plant, independent of any direct chemical effects the applied N may have on the availability of P sources in the soil. Among the growth promoting effects of N fertilization are increased root growth and favorable adjustments in plant metabolism, and the ability of unit areas of root surface to absorb P. Because soil P is considered immobile, gains in root volume per soil unit volume would improve utilization efficiency (53). Significant and economical growth responses have been obtained by applying P, or N plus P, to newly planted slash or loblolly pines on many lower coastal plain soils (71, 72). A commonly observed adverse biological effect is restriction of root development by N which, in turn, causes mortality or reduced top growth. Excessive soluble salts are particularly damaging on sandy soils and, according to Pritchett and Robertson (70), concentrations equivalent to 300 kg/ha or more of N or potash (K_2O) applied to southeastern coarse textured soils will reduce growth when application is made within the year of transplanting. Similarly, root suppression and death of slash pine seedlings have been associated with the addition of sodium salts added as $NaNO_3$ (49). This response often explains reduction in P uptake by N-fertilized seedlings grown in soils adequately supplied with P.

In coniferous species, uptake and the requirement of
nutrients may be strongly interrelated and correlated with
biomass production. For example, requirements for Ca and Mg
appear closely related to growth rate. Levels of these and
other essential elements must be considered when N and P
fertilization is prescribed (53). The demand for elements
increases as others are made available at higher levels for
plant growth. Following N fertilization, P often becomes
noticeable deficient because N has increased P demand. On
poorly drained soils of the lower coastal plain, N applied
without P suppressed slash pine growth. Rock phosphate or
superphosphate applied alone significantly increased diameter
and height growth 3 years after treatment, but the addition of
N with P further improved growth (69).

13.4.2 Site preparation and fertilization interactions.
Harvesting and reforestation practices can alter the nutrients
available during seedling establishment. In recent years,
considerable attention has been given to the potential for nutrient
losses to whole-tree harvesting and logging residue utilization.
The added biomass obtained through whole tree harvesting and
residue utilization clearly results in a greater proportional
nutrient loss than simply removing the bole. The significance
of such losses over the rotational life of the forest can be
known only by determining the nutrient input to a given site.
Intensive harvesting can change various aspects of the nutrient
cycle. Furthermore, prescribed burning can have adverse
effects on nutrient pools in pine ecosystems. Burning could
have a serious impact on ion transport because it directly
affects (a) uptake and return of elements by the forest, (b)
the nutrient capital of the soil system and (c) the mobility of
mineral elements within the soil system (32). Significant
quantities of cations and P, in addition to N, may be lost from
the forest floor during prescribed burning. On marginal sites
where inherent fertility may limit productivity, repeated
burning at short intervals could result in decreased productivity
(46). In weighing the effect of slash burning on soil properties,
consideration must be given to the severity of the burn. Light
fires do not remove all litter from the soil surface. Severe

fire removes all organic litter from the ground surface and bake the mineral soil to a crusty state. Light burning stimulates nitrification, but severe burns reduce N content. In addition, the effects of slash burning vary with different soils and locations within forest regions (88).

Mechanical site preparation can cause nutrient losses and results in soil physical disturbances. Harvesting and subsequent site preparation can markedly increase bulk density and decrease aeration (30), reducing root development and nutrient extraction. Sands and Bowen (77) determined that increasing the bulk density of soil from 1.35 g/cm³ to 1.60 g/cm³ decreased the length of radiata pine main root axis, first and second order laterals, and the diameter of first order laterals and the main axis. Windrowing removes top soil rich in nutrients and compacts the surface soil. Displacement can be minimized by careful rake positioning, shortening the distances between windrows and decreasing their size. Where sites can be prepared by chopping, burning or herbicide application, windrowing should be avoided (56).

On wet coastal plain soils, bedding generally improves seedling nutrition because it reduces vegetative competition, and improves organic matter concentration, drainage and aeration (82). Amelioration of the rooting environment during winter profoundly influences nutrient utilization and early growth of southern pines (54). Bedding clearly improves fertilization response (48). The effects of fertilization and water table control were additive in improving growth of slash and loblolly pines planted on strongly acid soil low in nutrients (100). When applied during bedding, fertilizer is incorporated into the soil where it is more readily available to young seedling root systems. The addition of N without bedding may suppress growth.

13.4.3. Moisture Stress and Weed Control. Tree growth response depends not only upon the form of fertilizer applied, but also on the amount of precipitation. Bengston and Voigt (10) reported that readily available forms were most efficient under moderate precipitation levels, while slowly soluble forms were superior under conditions of high precipitation. The availability of moisture impacts the response to fertilization. As conceptualized by Hansen (36), plant response to nutrient availability increases as moisture increases, then declines as soil aeration limits response (15) (Fig. 9). Fertilization can decrease yield if moisture is severely limiting (15). This is best illustrated with corn (Fig. 10). Nitrogen fertilization increased yield when precipitation was adequate but depressed production during drought. Red pine responded similarly (81). When moisture was inadequate, fertilization caused little response and salt toxicity increased (Fig. 11). Fertilization

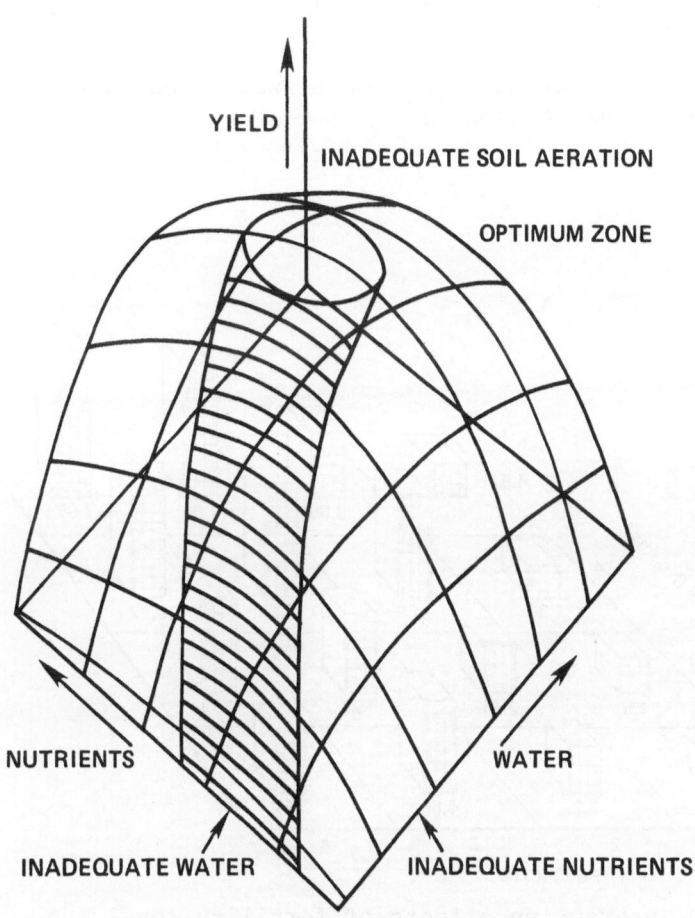

FIGURE 9. Diagramatic representation of interactions among fertilization, moisture availability and yield (36).

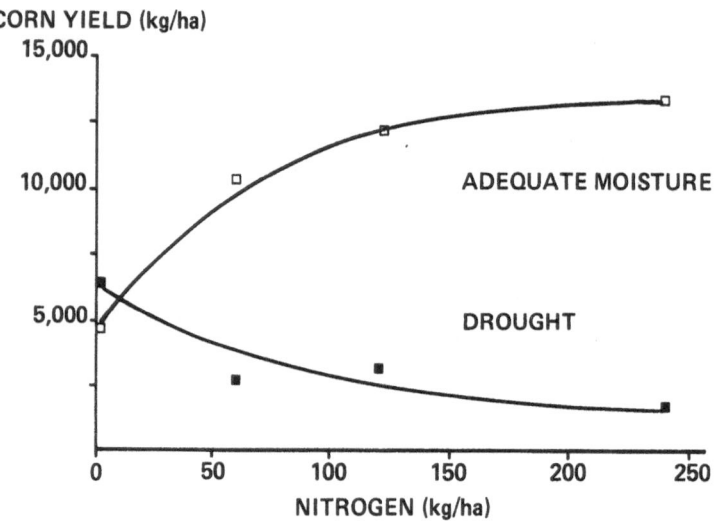

FIGURE 10. Effect of nitrogen fertilization on corn yield under two moisture regimes (15).

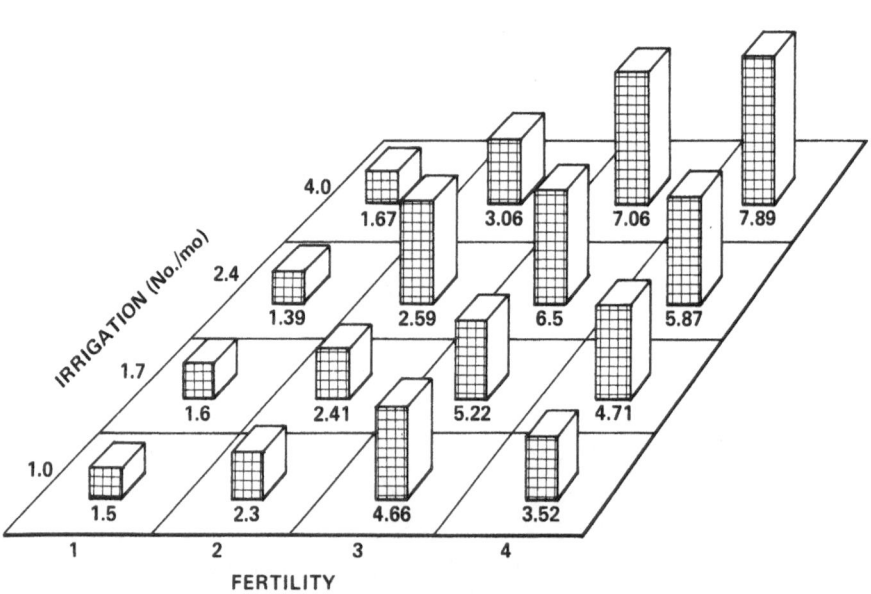

FIGURE 11. Interactive effects of fertility level and irrigation frequency on dry weight (g) of red pine seedlings (adapted from 81).

can alter growth in some situations by altering the plant's water status. During mild drought, fertilization can increase needle water potential and allow for greater potential growth (39). Fertilization can alter soil water potential as well as plant tissue water potential. Salt index is a relative measure of decreased soil solute potential caused by a specific fertilizer (Table 2). Referenced against $NaNO_3$, urea results in considerably less reduction in soil water potential. Sodium nitrate produces 3.7 times as much salt in the soil solution as does urea, per unit N added (71, 72). Diammonium phosphate (DAP) is the best combination fertilizer with a relative salt index of 0.07. Although osmotic troubles associated with high salts or low soil moisture may occur, Allen and Maki (1) reported survival of longleaf pine seedlings after drought was enhanced by a complete fertilizer, but not by N alone. Using the "yield after drought" criterion, Pharis and Kramer (65) reported too much or too little N intensified drought damage and decreased post-drought recovery. The danger of suppressing growth by excessive fertilizer salts can be reduced by deep localized applications, or by delaying fertilization for a year or more after transplanting (70).

Table 2. Salt indices of common fertilizers. Salt index is an indication of the relative decrease in solute potential produced by the dissolution of the fertilizer relative to that of the same weight of sodium nitrate (73).

Fertilizer	Analysis	Salt Index	Salt Index/ Nitrogen Unit
$NaNO_3$	16.5-0-0	100	1.00
NH_4NO_3	34-0-0	102	0.48
$(NH_4)_2SO_4$	01-0-0	69	0.54
Urea	45-0-0	73	0.26
TSP	0-45-0	10	0.04
DAP	18-46-0	29	0.07
KCL	0-0-60	116	0.32

Results have been inconclusive at times, but the conclusion is that fertilization without competition control during establishment is of little value to coniferous species (33, 35).

Fertilization without weed control increases competition for applied nutrients and available moisture. In addition, fertilization may increase competition from allelopathic plants which hampers seedling establishment and growth. As a result, absence of weed control often depresses growth and increases mortality. Without competition, N fertilization can have significant and long-lasting effects on early growth.

A possible alternative to competition control during establishment is use of slow-release fertilizers. These have been used successfully for fertilizing nurseries, ornamentals and Christmas trees. Agriform tablet (18-8-3) fertilization stimulated growth of loblolly pine seedlings and increased the root/shoot ratio. However, high rates caused poor root development, attributed to high salt concentrations near the roots (66). Slow-release fertilizers are well suited to reforestation because they can be added during planting, but high costs per unit N preclude widespread use (67).

13.4.4 Plant Growth Regulators. Some herbicides stimulate non-target plant growth beyond the benefits provided by weed control alone. There are many reports of increased plant yield (including tree species) at subherbicidal concentrations of triazines (27). The response of radiata pine to atrazine and simazine herbicides is a well documented example of growth stimulation. Sands and Zed (78) suggested that rates be increased to gain the added benefit of direct stimulation. These herbicides are now applied at twice the rate required for weed control in New Zealand and Australia. Atrazine application is particularly effective in stimulating growth soon after planting.

Numerous studies have shown the uptake and translocation of mineral elements by plants can be influenced by exogenously applied natural and synthetic plant growth regulators. Most work suggests that stimulation resulting from triazine application is associated with an increased efficiency in N uptake, particularly nitrate (27). Sands and Zed (78) determined atrazine stimulation of radiata pine was independent of the presence, amount or form (ammonium or nitrate) of N fertilizer.

Atrazine increased the concentrations and total contents of N,
P and K in radiata pine needles in both field and pot trials.

In addition to growth stimulation, herbicides can indirectly
affect seedling nutrition by influencing development of mycorrhizal fungi in soil and plant roots (60, 61, 83, 84). Douglas-
fir nursery seedlings treated with napropamide had significantly
more mycorrhiza types, reflecting a greater diversity of
mycorrhizal fungi, than those treated with DCPA or bifenox (91).

13.4.5. Genetic Influences and Interactions. Plant genotypes
differ in the uptake, translocation, accumulation and use of
mineral elements. Characteristics of genotypes influencing
uptake and utilization of mineral nutrients include: (a) ratio
of above-ground parts to roots, (b) physiological and biochemical processes and (c) morphological features of roots, stems
and leaves (80). These differences can be used to improve
plant growth under defined or constrained fertilizer conditions
(23).

With wide-scale planting of improved trees on relatively
infertile soil, an obvious development has been to seek genetic
lines that response well to added nutrients. Element-efficient
genotypes compared to inefficient plants produce more dry matter
per unit of the element provided (23). More specifically, the
desirable genotypes contain small concentrations of certain
mineral elements relative to the rate of biomass synthesis and
economic yield (79). Selections of trees that respond to
fertilization in a specific manner could: (a) markedly increase
the economic efficiency of fertilization, (b) increase the
value of tree improvement programs (25) and (c) alleviate
mineral stress without fertilization (79). However, selection
to produce fertilizer efficient trees must be done with a thorough understanding of seedling genotypic differences governing
mineral efficiency, utilization, and root exploitation of soil
nutrients as governed by absorptive capacity and root extension.

Stone (86) emphasized that large gains attributed to fertilizer actually result from interactions of nutrient supply with
other factors such as responsive genotypes and site preparation.
Therefore, information concerning interactions is of great
importance for efficient forest operations. Significant family

differences in absorption and efficiency suggest that soils from which provenances originate exert a strong selection pressure (89) and may require that nursery and planting site soil reactions be matched. Progeny have commonly shown differences in their ability to absorb major elements. Slash pine families, for example, differed in N and K absorption (98). The importance of genotype and fertilizer level interaction is apparent in the response of loblolly pine seedlings to five urea levels in field plantings (Fig. 12). During the first growing season, some of the 35 families tested grew well at all levels. Others grew poorly at both high and low application rates, but well at intermediate rates (76). The practical importance of genotype consideration is further supported by Bell et al. (8) who suggested Douglas-fir selection for growth based on efficient N fertilizer use might be successful.

Planting site diversity, and the potential for many rates and forms of operational fertilizers, cause considerable concern for the conditions and objectives of tree improvement programs. Variability within a single environment does not imply that selection will isolate genotypes with superior

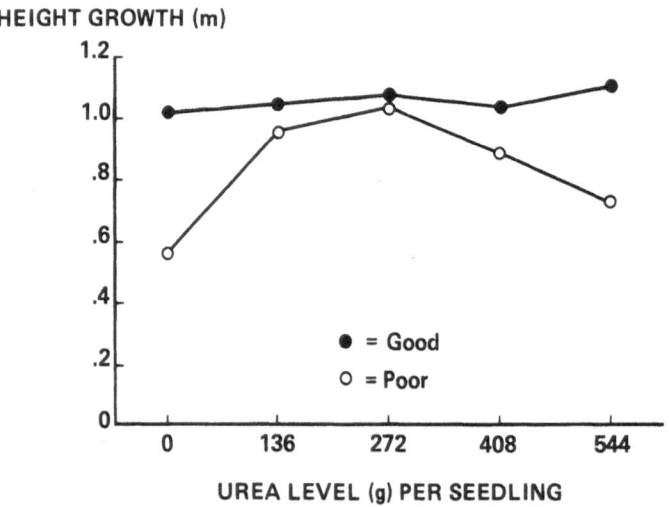

FIGURE 12. Growth of loblolly pine seedlings in response to urea where 'Good" = 5 best genotypes with no urea added and 'Poor' = 5 poorest genotypes with no urea added (76).

expression of the character over a range of environments or treatments. Likewise, there is no reason to believe any one physiological process will limit plant performance to the same extent in all species or environments (50). In this vein, Ballard (7) recommended progeny be tested under both fertilized and non-fertilized conditions to prevent inadvertent movement towards fertilizer-dependent populations in advanced generations. He also suggested the fertilizer rate and source correspond with that used operationally. For some species, such conditions appear unjustified and seriously challenge the profitability of maintaining seed orchards to capture specific genotype-nutrient synergisms. Matziris and Zobel (57) determined loblolly pine genotypes selected for good growth and form under non-fertilized conditions can usually be used under fertilized conditions with good performance expectations. It is, therefore, possible to breed broadly adapted, general utility strains of loblolly pine.

The proper management focus is not response to a specific treatment for the highest possible yield, regardless of the level of inputs. The ultimate goal is total stand performance and yield stability (24). A slow-growing family that responds exceptionally well to fertilization can produce less yield than a fast-growing family with a lesser response to fertilization. Also, some progeny lines grow well in unfertilized soil and, at the same time, respond well to fertilizer treatments. Such families appear to offer the greatest potential for improving growth (82).

13.5 ACKNOWLEDGEMENTS

The authors express their gratitude for support given by the New Mexico Agriculture Experiment Station and by the Mc-Intire-Stennis Forestry Research Program.

REFERENCES

1. Allen, R.M. and T.E. Maki. 1955. Response of longleaf pine seedlings to soils and fertilizers. Soil Sci. 79:359-362.

2. Anderson, H.W. and S.P. Gessel. 1966. Effects of nursery fertilization on outplanted Douglas-fir. J. For. 64:109-112.

3. Armson, K.A. and V. Sadreika. 1979. Forest tree nursery soil management and related practices. Nat. Res., Div. For., Ontario, Canada. 177 pp.

4. Autry, L.L. 1972. The residual effects of nursery fertilization and seed bed density levels on the growth of 12-, 14-, and 16- year old loblolly pine stands. M.S. Thesis. Miss. State Univ. 59 pp.

5. Baker, J.B. 1973. Intensive cultural practices increase growth of juvenile slash pine in Florida sandhills. For. Sci. 19:135-138.

6. Ballard, R. 1978. Use of fertilizers at establishment of exotic forest plantations in New Zealand. J. For. Sci. 8:70-104.

7. Ballard, R. 1980. The means to excellence through nutrient amendment, pp.159-200. In M.J. Wotton (ed.) Forest Plantations - The Shape of the Future (Vol.1). Proceedings of a symposium held at Tacoma, Wash., April 30-May 3, 1979.

8. Bell, H.E., R.F. Stettler and R.W. Stonecypher. 1979. Family X fertilizer interaction in one-year-old Douglas-fir. Silvae Genetica 28:1-5.

9. Bengston, G.W. 1976. Comparative response of four southern pine species to fertilization: effects of P, NP, and NPKMgS applied at planting. For. Sci. 22:487-494.

10. Bengston, G.W. and G.K. Voigt. 1962. A greenhouse study of relations between nutrient movement and conversion in a sandy soil and the nutrition of slash pine seedlings. Soil Sci. Soc. Am. Proc. 26:609-612.

11. Benzian, B. 1965. Experiments on nutrition problems in forest nurseries. For. Comm. Bull. 1:251 pp.

12. Benzian, B., R.M. Brown and C.R. Freeman. 1974. Effect of the late-season top-dressings of N(and K) applied to conifer transplants in the nursery on their survival and growth on British forest sites. Forestry. 47:153-184.

13. Blackmore, D.G. and W.G. Corns. 1979. Lodgepole pine and white spruce establishment after glyphosate and fertilizer treatments of grassy cutover forest land. For. Chron. 55:102-105.

14. Blatt, C.R. 1964. Nitrogen-phosphorus relationships in the nutrition of pitch pine (Pinus rigida). Diss. Abstr. 25, 2121.

15. Boyer, J.S. 1983. Crops, stress, and genetic improvement. Ill. Res. 25:18-20.

16. Bradbury, I.K. and D.C. Malcolm. 1977. The effect of phosphorus and potassium on transpiration, leaf diffusive resistance and water-use efficiency on Sitka spruce seedlings. J. Appl. Ecol. 14:631-641.

17. Brix, H. and R. van den Driessche. 1974. Mineral nutrition of container-grown tree seedlings, pp.77-84. In R.W. Tinus, W.I. Stein and W.E. Balmer (eds.) Proc. N. Amer. Containerized For. Tree Seedling Symp., Denver, Colo.

295

18. Buckner, E. and T.E. Maki. 1977. Seven-year growth of fer-
    tilized and irrigated yellow-poplar, sweetgum, northern
    red oak, and loblolly pine planted on two sites. For. Sci.
    23:402-410.
19. Burns, R.M. and E.A. Hebb. 1972. Site preparation and
    reforestation of droughty, acid sands. USDA Agr. Hndbk No.
    426. 61 pp.
20. Carlson, W.C. and C.L. Preisig. 1980. Effects of
    controlled-released fertilizers on the shoot and root
    development of Douglas-fir seedlings. Can. J. For. Res.
    11:230-242.
21. Carlson, W.C. and C.L. Preisig 1981. Effects of
    controlled-release fertilizers on shoot and root
    development of outplanted western hemlock (Tsuga
    heterophylla) seedlings. Can. J. For. Res. 11:752-757.
22. Christersson, L. 1973. The effect of inorganic nutrients
    on water economy and hardiness on conifers. I. The
    effect of varying potassium, calicium, and magnesium
    levels on water content, transpiration rate, and the
    initial phase of development of frost hardiness of Pinus
    silvestris. Studia For. Suec. 103:5-26.
23. Clark, R.B. 1983. Plant genotype differences in the
    uptake, translocation, and use of mineral elements
    required for plant growth. Plant Soil. 72:175-196.
24. Dambroth, M. and E. Bassam. 1983. Low input varieties;
    definition, ecological requirements and selection. Plant
    Soil. 72:365-377.
25. Davey, C.B. 1968. Biological considerations in fertilizer
    evaluation, pp.264-274. In Forest Fertilization: Theory
    and Practice, Symp. For. Fert. 1967. Tenn. Valley
    Authority, Muscle Shoals, AL.
26. Dickson, D.A. and P.S. Savill. 1974. Early growth of
    Picea sitchensis (Bong.) Carr. on deep oligotropic peat
    in Northern Ireland. Forestry. 47:57-88.
27. Ebert, E. and S.W. Dumford. 1976. Effects of triazine
    herbicides on the physiology of plants. Residue Rev.
    65:1-103.
28. Edgren, J.W. 1966. Bed density and size of ponderosa
    pine seedlings at the Bend nursery. Repr. from Proceedings
    of 10th Biennial Western Forest Nursery Council Meeting.
29. Flinn, D.W., P. Hopmans, I. Moller and K. Tregonning.
    1979. Response of radiata pine to fertilizers containing
    N and P applied at planting. Aust. For. 42:125-131.
30. Gent, Jr., J.A., R. Ballard and A.E. Hassan. 1983. The
    impact of harvesting and site preparation on the physical
    properties of lower coastal plain forest soils. Soil
    Sci. Soc. Am. J. 47:595-598.
31. Glatzel, G. 1971. Fertilizer treatment in high altitude
    afforestation. Allg. Forstztg. 82:281-283. (Abst.)
32. Grier, C.C. and D.W. Cole. 1971. Influence of slash
    burning on ion transport in a forest soil. Northwest
    Sci. 45:100-106.
33. Gjerstad and L.R. Nelson. 1983. Accelerating early growth
    in plantations: Growth response and physiology of tree
    seedlings as affected by weed control. In New Forests for

a Changing World, Proc. SAF Nat. Conv., Oct. 16-20, 1983, Portland, OR. (In Press).

34. Grunes, D.L. 1959. Effect of nitrogen on the availability of soil and fertilizer phosphorous to plants. Adv. Agr. 11:369-396.

35. Hanschke, D. 1966. (Chemical weed control in the establishment and tending of plantations). Dissertation, Forstliche Fakulat der Georg-August-Universtat zu Gottingen, Hann, Munden. 134-155 pp.

36. Hansen, E.A. 1976. Determining moisture-nutrient requirements for maximum fiber yield, pp.43-46. In Intensive Plantation Culture. USDA For. Ser. Gen. Tech. Rep. NC-21.

37. Hauge, T.P. 1971. Root growth after planting and fertilizing Norway spruce on wet peatland: results of six growing seasons. Tidsskr. Skogbr. 79:315-326. (Abst.)

38. Hemminger, G.D. and H.D. Gerhold. 1969. Effects of organic fertilizer on Douglas-fir provenance study. Penn. State Univ. Res. Briefs. 4:33-36.

39. Hillerdal-Hagstromer, K., E. Mattson-Djos and J. Hellvist. 1982. Field studies of water relations and photosynthesis in Scots pine. II. Influence of irrigation and fertilization on needle water potential of young pine trees. Physiol. Plant. 54:295-301.

40. Hinesley, L.E. and T.E. Maki. 1980. Fall fertilization helps longleaf pine nursery stock. South. J. Appl. For. 4:132-135.

41. Hoffman, F. 1962. (Interactions of phosphorous and nitrogen in the nutrition of spruce seedlings. Zemlj. Bilj. 11:341-348). (Abst.)

42. Ingestad, T. 1977. Nitrogen and plant growth; maximum efficiency of nitrogen fertilizers. Ambio 6:146-151.

43. Ingestad, T. 1979. A definition on optimum nutrient requirements in birch seedlings. Physiol. Plant. 46:31-35.

44. Jurgensen, M.F. and A.L. Leaf. 1965. Soil moisture-fertility interactions related to growth and nutrient uptake of red pine. Soil Sci. Soc. Am. Proc. 29:294-299.

45. Kaufman, C.M., W.L. Pritchett and R.E. Choate. 1977. Growth of slash pine (Pinus elliotii Engel. var. elliottii) on drained flatwoods. Fla. Agr. Exp. Sta. Bull. 792, 30 pp.

46. Kodama, H.E. and D.H. Van Lear. 1980. Prescribed burning and nutrient cycling relationships in young loblolly pine plantations. South. J. Appl. For. 4:118-122.

47. Lopushinski, W. and T. Beebe. 1976. Relationship of shoot-root ratio to survival and growth of outplanted Douglas-fir and ponderosa pine seedlings. USDA For. Ser. Res. Note. PNW-274, 7 pp.

48. Malac, B.F. 1968. Research in forest fertilization at Union Camp Corp. pp.203-208, In Forest Fertilization: Theory and Practice. Tenn. Valley Auth., Muscle Shoals, Al.

49. Maftoun, M. and W.L. Pritchett. 1970. Effects of added nitrogen on the availability of phosphorus to slash pine on two lower coastal plain soils. Soil Sci. Soc. Am. Proc. 34:685-690.

50. Mahon, J.D. 1983. Limitations to the use of physiological variability in plant breeding. Can. J. Plant Sci. 63:11-21.

51. Matziris, D. and B.J. Zobel. 1976. Effect of fertilization on growth and quality characteristics of loblolly pine. For. Ecol. Manag. 1:21-30.

52. McFee, W.W. and E.L. Stone. 1968. Ammonium and nitrate as nitrogen sources for Pinus radiata and Picea glauca. Soil Sci. Soc. Am. Proc. 32:879-884.

53. McKee, Jr., W.H. 1973. Slash pine response to nitrogen and phosphorus on imperfectly drained soil of the West Gulf Coastal Plain. Soil Sci. Soc. Am. Proc. 37:784-788.

54. McKee, W.H. and E. Shoulders. 1970. Depth of water table and redox potential of soil affect slash pine growth. For. Sci. 16:399-402.

55. Mexal, J.G. 1980. Seedling bed density influences yield and performance, pp.89-95. In Proc. of 1980 South. Nur. Conf., Lake Barkley, KY. USDA Tech. Pub. SA-TP17.

56. Morris, L.A., W.L. Pritchett and B.F. Swindel. 1983. Displacement of nutrients into windows during site prepartion of a flatwood forest. Soil Sci. Soc. Am. J. 47:591-593.

57. Morrison, I.K. 1974. Mineral nutrition of conifers with special reference to nutrient status interpretation: a review of the literature. Dept. of Environment. Can. For. Ser. Publ. 1343, 74 pp.

58. Moschler, W.W., G.D. Jones and R.E. Adams. 1970. Effects of loblolly pine fertilization on a piedmont soil: growth, foliar composition, and soil nutrients ten years after establishment. Soil Sci. Soc. Am. Proc. 34:683-685.

59. Mullin, R.E. 1981. Seedbed density effects on red pine in the nursery and after outplanting. Ontario Min. Natur. Resour., Nursery Notes No.72, 5 pp.

60. Nemec, S. and D. Tucker. Effects of herbicides on endomycorrhizal fungi in Florida citrus (Citrus spp.) soils. Weed Sci. 31:427-431.

61. Palmer, Jr., J.G.. J.E. Kuntz and J.G. Palmer, Sr. 1979. Effects of the herbicide dacthal on mycorrhizal development of Pinus resinosis Ait. in southern Wisconsin J. Arbor. 5:162.

62. Pellett, H.M. and Cater, J.V. 1981. Effects of nutritional factors on cold hardiness of plants. Hort. Rev. 3:144-171.

63. Perry, T.O. 1971. Dormancy of trees in winter. Science 171:29-36.

64. Pharis, R.P., R.L. Barnes and A.W. Naylor. 1964. The effects of nitrogen level, calcium level, and nitrogen source upon the growth and composition of Pinus taeda L. Physiol. Plant. 17:560-572.

65. Pharis, R.P., P.J. Kramer. 1964. The effects of nitrogen and drought on loblolly pine seedlings. 1. Growth and composition. For. Sci. 10:143-150.

66. Pope, P.E. and J.E. Voeller. 1976. Fertilization effects on survival and early growth of planted loblolly pine seedlings in the Ozarks. Ark. Farm Res. 25:13.

67. Pritchett, W.L. 1979. Properties and management of forest soils. John Wiley and Sons, New York, 500 pp.

68. Pritchett, W.L. 1979. Site preparation and fertilization on slash pine on a wet savanna soil. South. J. Appl. For. 3:86-90.

69. Pritchett, W.L. and W.R. Llewellyn. 1966. Response of slash pine to phosphorus in sandy soils. Soil Sci. Soc. Am. Proc. 30:509-512.

70. Prichett, W.L. and W.L. Robertson. 1960. Problems relating to research in forest fertilization with southern pines. Soil Sci. Soc. Amer. Proc. 24:510-512.

71. Pritchett, W.L. and W.H. Smith. 1970. Fertilizing slash pine on sandy soils in lower coastal plain, pp.19-41. In Proc. 3rd North Amer. For. Soils Conf., Oregon State Univ., Corvallis, Ore.

72. Pritchett, W.L. and W.H. Smith. 1974. Management of wet savanna forest soils for pine production. Fla. Agr. Exp. Sta. Tech. Bull. 762, 22 pp.

73. Rader, Jr., L.M. White and C.W. Whittaker. 1943. The salt index: A measure of the effect of fertilizers on the concentration of the soil solution. Soil Sci. 55:201.

74. Richards, B.N. and D.I. Bevege. 1967. Effect of cultivation and fertilization on potential yield of pulpwood from loblolly pine. Aust. For. 31:202-210.

75. Richards, B.N. and G.L. Wilson. 1963. Nutrient supply and mycorrhiza development in Carribean pine. For. Sci. 9:405-412.

76. Roberds, J.H., G. Namkoong and C.B. Davey. 1976. Family variation in growth response of loblolly pine to fertilizing with urea. For. Sci. 22:291-299.

77. Sands, R. and G.D. Bowen. 1978. Compaction of sandy soil in radiata pine forest. II. Effects of compaction on root configuration and growth of radiata pine seedlings. Aust. J. Soil Res. 17:101-113.

78. Sands, R. and P.G. Zed. 1979. Promotion of nutrient uptake and growth of radiata pine by atrazine. Aust. For. Res. 9:101-110.

79. Saric, M.R. 1981. Genetic specificity in relation to plant mineral nutrition. J. Plant Nutrition 3:743-766.

80. Saric, M.R. 1983. Theoretical and practical approaches to the genetic specificity of mineral nutrition of plants. Plant Soil 72:137-150.

81. Schomaker, C.E. 1969. Growth and foliar nutrition of white pine seedlings as influenced by simultaneous changes in moisture and nutrient supply. Soil Sci. Soc. Amer. Proc. 33:614-618.

82. Schultz, R.P. 1975. Managing genetically improved southern pines, pp.209-237. In B.A. Thielges (ed.) Forest Tree Improvement - The Third Decade. 24th Annu. For. Symp. Louisiana State Univ., Baton Rouge, La.

83. Smith, J.R. and B.W. Ferry. 1979. The effects of simazine applied for weed control on the mycorrhizal development of Pinus seedlings. Ann. Bot. 43:93-99.

84. South, D. 1981. Effects of selected herbicides on endomycorrhizal formation of sweetgum (Liquidambar styraciflua L.). Fifth N. Amer. Conf. on Mycorrhizae, Quebec. (Abst.).

85. Stoeckler, J.H. and H.F. Arnemen. 1960. Fertilizers in Forestry. Adv. Agr. 12:127-195.

86. Stone, E.L. 1973. Biological objectives in forest fertilization, pp.10-18. In Proc. Forest Fertilization Symposium NE For. Exp. Stn., Upper Darby, Penn. (USDA For. Serv., Gen. Tech. Rep. NE-3).

87. Switzer, G.L. and L.E. Nelson. 1963. Effects of nursery fertility and density on seedling characteristics, yield and field performance of loblolly pine. Soil Sci. Soc. Am. Proc. 27:461-464.

88. Tarrant, R.F. 1956. Effects of slash burning on some soils of the Douglas-fir region. Soil Sci. Soc. Am. Proc. 20:408-411.

89. Tiessier du Cros, E. and B. Lepoutre. 1983. Soil X provenance interaction in beech (Fagus silvatica L.). For. Sci. 29:403-311.

90. Timmis, R. 1974. Effect of nutrient stress on growth, bud set and hardiness in Douglas-fir seedlings, pp.187-193. In R.W. Tinus, W.I. Stein and W.E. Balmer (eds.) Proc. North Amer. Containerized For. Tree Seedlings Symp., Denver, Co. . Great Plains Agr. Counc. Publ. 68.

91. Trappe, J.M. 1983. Effects of herbicides bifenox, DCPA, and napropamide on mycorrhiza development of ponderosa pine and Douglas-fir seedlings in six western nurseries. For. Sci. 29:464-468.

92. Tukey, H.B. and M.M. Meyer. 1966. Nutrient applications during the dormant season. Proc. Int. Plant Propag. Soc. 16:306-310.

93. van den Driessche, R. 1977. Fertilizer experiments in conifer nurseries of British Columbia. B.C. For. Ser. Res. Note No. 79, 32 pp.

94. van den Driessche, R. 1980. Health, vigour and quality of conifer seedlings in relation to nursery soil fertility, pp.100-120. In Proc. N. Amer. For. Tree Nursery Soils Workshop, July 28-Aug. 1, 1980, Syracuse, New York.

95. van den Driessche, R. 1982. Relationship between spacing and nitrogen fertilization of seedlings in the nursery, seedling size, and outplanting performance. Can. J. For. Res. 12:865-875.

96. Wakeley, P.C. 1969. Results of southern pine planting experiments established in the middle twenties. J. For. 67:237-241.

97. Walker, L.C. 1962. The effects of water and fertilizer on loblolly and slash pine seedlings. Soil Sci. Soc. Am. Proc. 26:197-200.

98. Walker, L.C. and R.D. Hatcher. 1965. Variation in the ability of slash pine progeny groups to absorb nutrients. Soil Sci. Soc. Am. Proc. 29:616-621.

99. Weetman, G.F. 1962. Establishment report on a humus decomposition experiment. Woodlands Research Index No. 134, Pulp Pap. Res. Inst. Can., Montreal.

100. White, E.H. and W.L. Pritchett. 1970. Water table control and fertilization for pine production in the flatwoods. Fla. Agr. Exp. Sta. Bull. 743, 41 pp.

14. MYCORRHIZAE AND REFORESTATION SUCCESS IN THE OAK-HICKORY REGION

R.K. DIXON, H.E. GARRETT, G.S. COX, and S.G. PALLARDY
Assistant Professor, Department of Forest Resources,
University of Minnesota, St. Paul, MN 55108, Professors
and Assistant Professor, School of Forestry, Fisheries,
and Wildlife, University of Missouri, Columbia, MO 65211.

ABSTRACT

The potential for improving the growth of outplanted oak
seedlings through the manipulation of mycorrhizal symbiosis is
promising. However, only limited research has been conducted
on the mycorrhizal relationships of commercially valuable oak
species. This study was conducted to evaluate the role of
mycorrhizal inoculation in an oak reforestation program.

Black oak (Quercus velutina Lam.) seedlings inoculated
with Pisolithus tinctorius (Pers.) Coker and Couch were grown
for one season in Spencer-LeMaire containers in a glasshouse
and in a nursery; noninoculated seedlings were also grown in
containers and in the nursery for comparison. Examination of
the inoculated stock before outplanting revealed that approximately 40 percent of the roots of the containerized and 1-0
bareroot seedlings were infected with P. tinctorius. Bareroot seedlings were significantly larger than container-
grown seedlings. However, total root system length of the
containerized seedlings was significantly greater.

Two years after outplanting on two Missouri Ozark clearcut
sites, shoot length, root collar diameter, and leaf area of
the container-grown seedlings inoculated with P. tinctorius
were greater than in the inoculated and noninoculated bareroot stock. The bareroot stock, regardless of inoculation
treatment, suffered from repeated shoot dieback, and leaf area
remained relatively small. Ectomycorrhizal infection of
roots of the container-grown stock by P. tinctorius was

Duryea, M.L. and Brown, G.N. (eds.). Seedling physiology and reforestation succes
©1984, Martinus Nijhoff/Dr W. Junk Publishers, Dordrecht/Boston/London. ISBN 978-94-009-6139-5

22 percent after one year in the field. During the same
period of time ectomycorrhizal infection of the bareroot
stock declined to 16 percent. Total root system length was
significantly greater for P. tinctorius inoculated container-
grown stock.

Seasonal evaluation of seedling water relations indicated
that containerized stock inoculated with P. tinctorius
avoided water stress more effectively during a mild drought.
Soil-plant liquid flow resistance was significantly lower in
P. tinctorius inoculated seedlings than in noninoculated
stock.

14.1 INTRODUCTION

Oaks are the most widely distributed hardwood trees in
North America. In the eastern U.S., the oak-hickory (Quercus-
Carya) and oak-pine (Pinus) associations cover approximately
150 million acres (31). Despite their widespread occurrence
and the problems associated with natural regeneration,
artificial regeneration of oaks has not been widely practiced
(19,20). There is, however, a need for artificial regeneration wherever cutover areas are poorly stocked (2,39), where
natural regeneration fails (37) or when genetically superior
stock is desired (21). An increasing awareness of the
problem of obtaining adequate oak regeneration has stimulated
interest in oak planting (18,21).

Conventional planting of bareroot and container-grown
black (Q. velutina Lam.), northern red (Q. rubra L.) and
white (Q. alba L.) oak seedlings has met with only limited
success (19,20,29). Survival of oak planting stock is generally acceptable but early growth is slow. Successful
artificial regeneration of oak requires that seedlings possess
a minimum shoot growth potential of 0.5 m per year following
outplanting. Unfortunately, even on the best sites, shoot
growth of newly established seedlings seldom averages more
than a few centimeters per year. Juvenile oak characteris-
tically suffer from periodic shoot dieback and substantial
reduction in leaf area (18,20). The poor performance of

303

planted oaks may be partially attributed to inadequate root:
shoot ratios and poor physiological condition of stock prior
to outplanting (21).

The inoculation of oak seedlings with selected ectomycorrhizal fungi may result in physiologically superior planting
stock with a more functional and responsive root system (13,
14,40). Mycorrhizal relationships of forest trees, including
species in the genus Quercus, are involved in nutrient uptake
(14,15), water absorption (8,32), protection against selected
root pathogens (22) and in tolerance of environmental stress
(14). Although descriptive information on mycorrhizal
relationships of oak has been published, only recently has the
significance of specific ectomycorrhizae to seedling growth
and development been recognized (13). Previous investigations
with pine planted over a range of sites have shown that
survival and growth of bareroot and container-grown seedlings
are significantly improved following inoculation with the
ectomycorrhizal fungus P. tinctorius (26,27,28). Recent
preliminary investigations with northern red and white oak
indicate that moderate P. tinctorius ectomycorrhizal
development significantly stimulates seedling growth (23,24)
and drought resistance (8).

The objective of this study was to compare and characterize
the early growth and water relations of P. tinctorius inoculated and noninoculated container-grown and 1-0 bareroot
black oak seedlings outplanted in an upland clearcut site
in the Missouri Ozark region.

14.2 MATERIALS AND METHODS
14.2.1 Glasshouse seedling production procedures

Prior to the initiation of seedling production, vegetative
inoculum of P. tinctorius was grown in a vermiculite-peat moss-
nutrient mixture for three months using methods described by
Marx and Bryan (25). Container-grown seedlings were produced
in a glasshouse at the University of Missouri-Columbia using
procedures described by Tinus and McDonald (42). Acorns,
collected from a mid-Missouri source, were sown in

Spencer-LeMaire Super 45 bookplanters. All cavities were filled with a 1:1 homogeneous mixture of sterile peat moss and construction grade vermiculite. Treatments consisted of P. tinctorius inoculated and noninoculated seedlings. One-half of the bookplanters were inoculated with P. tinctorius at the time of acorn planting by thoroughly mixing 35 cm^3 of vegetative mycelium (1:22 by volume) into the growth medium using procedures described by Riffle and Maronek (33). Beginning two weeks after acorn germination, seedlings were fertilized bi-weekly with a complete nutrient solution (16). Daily seedling photoperiod was approximately 15 hours, and ambient glasshouse temperature ranged from 21-33°C. Seedlings were watered as needed.

14.2.2 Nursery seedling production procedures

The bareroot seedlings were grown at the George O. White State Forest Nursery, Licking, MO. Nursery beds were tilled and fumigated under polyethylene with 740 kg/ha of Dowfume MC-2 (Dow Chemical Company, Midland, MI). Soil chemical analyses from all plots before the experiment were similar: pH 6.2, and 63, 224, 11, 154, 803, and 1 ppm of available P, K, NO_3, Mg, Ca, and Zn, respectively. The soil texture was a sandy loam and organic matter content was approximately 2% in all beds.

Treatments consisted of P. tinctorius inoculated and non-inoculated seedlings. Randomly selected plots within the nursery beds were inoculated with vegetative inoculum of P. tinctorius. The vegetative mycelium-peat moss-vermiculite mixture was broadcast over the nursery beds at a concentration of 2.2 L/m^2 of soil surface and immediately mixed into the upper 20 cm of soil. Control plots received an equal volume of sterile inoculum in the same manner. Acorns were row planted to obtain approximately 100 seedlings/m^2 in all plots. The seedlings did not receive any supplemental fertilizers during the growing season and were irrigated as needed. Routine cultural practices were utilized when possible.

Both nursery- and glasshouse-grown seedlings were harvested five months after acorn sowing. In the nursery, plots were undercut with a root pruning bar at a depth of 36 cm and the seedlings lifted by hand. A sample of inoculated and non-inoculated nursery- and container-grown seedlings were randomly chosen for measurements of shoot length, leaf surface area, root system length, and ectomycorrhizal development. Criteria for identifying and quantifying ectomycorrhizal infection were described by Dixon et al. (5). The remaining bareroot seedlings were stored at 5°C in moist peat moss in Kraft paper bags; container-grown stock were moved to cool, shaded conditions prior to outplanting.

14.2.3 Site conditions and seedling outplanting procedures

Two outplanting sites were selected on the Sinkin Experimental Forest within the Mark Twain National Forest in Dent County, MO. The two 0.4 ha sites, which were composed of mature, mixed stands of oak, hickory and shortleaf pine (P. echinata Mill.), were clearcut. Debris was windrowed and the herbicide 2,4,5-T was applied at a concentration of 3 kg of active ingredient (acid equivalent) in 70 L of water per hectare. Each site had a 5-20% slope and a northwest aspect. Site index for black oak on both planting sites was 23 m at a base age of 50 years. The soils on both sites were cherty silty clay loams of the Wilderness and Poyner series, containing 1-2% organic matter and 567, 10, 40, 180, and 45 ppm of N, P, K, Ca, and Mg, respectively, in the upper 50 cm of the soil profile. Soil pH ranged from 4.1 to 4.3 on both sites.

The seedlings were hand planted using mattocks in the fall of 1978. The outplanting was arranged in randomized complete blocks on each site. The four treatments, (i) container-grown inoculated, (ii) container-grown noninoculated, (iii) bareroot inoculated, and (iv) bareroot noninoculated, were randomly assigned to plots within blocks. Seedlings were planted at 1 x 1 meter spacing within blocks.

14.2.4 Evaluation of seedling field performance

Seedling shoot length, leaf surface area, root system length, and ectomycorrhizal development were evaluated one and two years after outplanting. Leaf area was determined using a portable area meter (LI-COR model LI-3000) and root system length using the line intersection technique (34). Seedling leaf water potential (Ψ, MPa), leaf conductance (g_1, $cm \cdot s^{-1}$), leaf temperature (LT, °C), ambient air temperature (°C), and relative humidity (RH, %) were evaluated weekly within each treatment during the first growing season. Seedling sampling procedures, instrumentation and techniques were described in detail by Dixon et al. (7). Seedling growth and water relations data were subjected to analyses of variance, and differences between treatment means were tested by the least significant difference test.

14.3 RESULTS

14.3.1 Seedling characteristics prior to outplanting

Pisolithus tinctorius formed ectomycorrhizae on approximately 40% of the lateral roots of seedlings in containers and the nursery prior to outplanting (Table 1). However, because there were more than twice as many lateral roots on container-grown stock, the aggregate number of primary laterals infected with P. tinctorius was 130% greater among containerized seedlings than among nursery-grown seedlings. Furthermore, all container-grown seedlings originally inoculated with P. tinctorius developed extensive yellow-gold hyphal strands which permeated the peat-vermiculite growth medium. In the noninoculated control treatments, significantly fewer P. tinctorius ectomycorrhizae were observed in the nursery and no ectomycorrhizal structures were observed among the containerized seedlings.

Seedling leaf area was the only growth variable significantly increased in the nursery or glasshouse following inoculation with P. tinctorius (Table 1). Shoot length and leaf area of the bareroot stock were significantly larger than for the container-grown seedlings. In contrast, total root system

Table 1. Growth characteristics of Pisolithus tinctorius inoculated and noninoculated container-grown and 1-0 bareroot black oak seedlings before and after outplanting in the Sinkin Experimental Forest (6,9).

Treatment	Shoot length (cm)		Leaf area (cm^2)		Root system length (cm)		Ectomycorrhizal laterals (%)	
	pre-plant	year 1	pre-plant	year 1	pre-plant	year 1	pre-plant	year 1
Container-grown Inoculated	12.6b[1]	18.6a	203c	278a	1142a	1051a	39a	22a
Container-grown Noninoculated	12.1b	16.0ab	175d	179ab	944a	798b	0c	1c
1-0 Bareroot Inoculated	22.0a	14.7b	564a	126b	530b	162c	38a	16b
1-0 Bareroot Noninoculated	18.9a	12.9b	491b	27c	347b	210c	25b	2c

[1]/Within a column (pre-plant or year 1), growth variable means not followed by a common letter are significantly different by the LSD test ($P \leq 0.05$).

length of the container-grown seedlings was twice as large as that of the bareroot stock before outplanting.

14.3.2 Seedling characteristics one year after outplanting

Shoot length, leaf area, and root system length of the container-grown seedlings inoculated with P. tinctorius were significantly larger than the bareroot stock regardless of inoculation treatment, one year after outplanting (Table 1). Root system length and leaf area of the P. tinctorius inoculated container-grown seedlings were approximately 5x and 2x greater, respectively, than that of inoculated bareroot seedlings. Abundance of Pisolithus tinctorius ectomycorrhizae declined on the root systems of both stock types after one year in the field. The decline in the numbers of ectomycorrhizae was substantially greater on the bareroot seedlings compared to the containerized stock.

14.3.3. Water relations of seedlings one year after outplanting

Throughout the study, precipitation was frequent and soil water potential within the rooting zone remained higher than −0.40 MPa (Dixon et al. 1983). Over the course of the growing season, Ψ_{pd} (predawn) of the container-grown stock was slightly higher than that of the bareroot seedlings (Table 2). Seasonal averages of Ψ_{pd} indicates no effect of ectomycorrhizal inoculation with P. tinctorius. However, during a short drought the container-grown seedlings inoculated with P. tinctorius exhibited significantly higher Ψ_{pd} values when compared to other stock types. Containerization and inoculation with P. tinctorius both significantly improved seedling capacity to moderate depression in Ψ_{sn} (solar noon) and reduced the absolute difference between Ψ_{pd} and Ψ_{sn}. Soil-plant liquid flow resistance (R_{s-p}) was significantly lower in plants grown in containers or inoculated with P. tinctorius before outplanting.

Table 2. Mean values of water relations characteristics of
Pisolithus _tinctorius_ inoculated and noninoculated
container-grown and 1-0 bareroot black oak seedlings
one year after outplanting in the Sinkin Experimental
Forest (7).

Treatment	Mild drought Ψ_{pd} (-MPa)	Ψ_{pd} (-MPa)	Whole season $\|\Psi_{sn} - \Psi_{pd}\|$ (MPa)	R_{s-p} [2] $(MPa \cdot cm^2 \cdot s \cdot \mu g\ H_2O^{-1})$
Container-grown Inoculated	0.20a[1]	0.08a	1.16a	4.52a
Container-grown Noninoculated	0.30b	0.10ab	1.31a	5.05b
Bareroot Inoculated	0.35b	0.15ab	1.68b	5.24c
Bareroot Non-inoculated	0.53c	0.17b	1.98c	7.88d

[1] Means within a column not followed by a common letter are
significantly different by the LSD test ($P \leq 0.05$).

[2] $R_{s-p} = |\Psi_{pd} - \Psi_{sn}|/\log(TFD + 1)$, where TFD is transpirational
flux density ($\mu g\ H_2O \cdot cm^{-2} \cdot s^{-1}$).

14.3.4 Seedling characteristics two years after outplanting

Shoot length, root collar diameter, and leaf surface area
were significantly improved by containerization and _P._
tinctorius inoculation two years after outplanting (Table 3).
The bareroot seedlings, regardless of inoculation treatment,
exhibited negative shoot growth. Repeated shoot dieback,
coupled with relatively poor root growth (Table 1) was a
common characteristic of the bareroot stock.

Table 3. Growth characteristics of <u>Pisolithus</u> <u>tinctorius</u>
inoculated and noninoculated container-grown and
1-0 bareroot black oak seedlings two years after
outplanting in the Sinkin Experimental Forest.

Treatment	Shoot length (cm)	Root collar diameter (cm)	Leaf area (cm^2)
Container-grown Inoculated	28a[1/]	0.75a	650a
Container-grown Noninoculated	21b	0.64b	488b
1-0 Bareroot Inoculated	14c	0.56bc	409bc
1-0 Bareroot Noninoculated	12c	0.52c	323c

[1/] Within a column, growth variable means not followed by a
common letter are significantly different by the LSD test
(p $\leq$ 0.05).

14.4 DISCUSSION

14.4.1 <u>Seedling inoculation success in the glasshouse and nursery</u>

Inoculation of black oak seedlings with the ectomycorrhizal
fungus <u>P</u>. <u>tinctorius</u> was successful in both nursery beds and
containers in a glasshouse. Results of this study suggest that
few changes in routine nursery or glasshouse cultural practices
are needed to produce oak seedlings with <u>P</u>. <u>tinctorius</u> ectomycorrhizae. The current practice in nursery and glasshouse
operations of frequently saturating oak seedling root systems
with concentrated doses of fertilizers to produce seedlings of
a plantable size may discourage significant ectomycorrhizal
development (5,10,35). In this study, relatively low soil
fertility regimes resulted in successful inoculation of seedlings
which stimulated a significant growth response. Therefore,
frequent heavy applications of soil fertilizer may be
unnecessary and even undesirable for the production of ectomycorrhizal oak seedlings for outplanting. Previously, Dixon
et al. (5) demonstrated that reduced soil fertility improved
the growth of <u>P</u>. <u>tinctorius</u> inoculated black oak seedlings.

The results indicate that vegetative inoculum of P. tinctorius is a suitable form of inoculum for early development of ectomycorrhizae on black oak. Following soil sterilization, thorough mixing of inoculum in either container growth medium or the upper 20 cm of nursery soil resulted in P. tinctorius colonization of primary laterals throughout the root zone. Pisolithus tinctorius successfully competed with other species of naturally occurring soil fungi, colonized the root systems of black oak, and stimulated a significant seedling growth response. In container culture, no ectomycorrhizae were observed on the noninoculated stock indicating windborn inoculation is not automatic and deliberate inoculation is sometimes necessary (42). These results suggest that artificial inoculation techniques result in early ectomycorrhizal development and are useful in improving oak seedling growth.

The early development of ectomycorrhizae by P. tinctorius in nursery soils and in containers was associated with significant improvements in seedling growth prior to outplanting. Similarly, Marx (23,24) reported that northern red and white oak seedlings inoculated with P. tinctorius were significantly larger than noninoculated controls. Bjeckford et al. (1), working with container-grown northern red oak, observed that seedlings inoculated with P. tinctorius formed relatively small amounts of ectomycorrhizae and performed little better than noninoculated controls. Past work by the authors (4) demonstrated relatively poor ectomycorrhizal colonization and subsequent growth of container-grown black oak seedlings if environmental conditions were unfavorable. Interactions between the fungus and host are complex, and environmental conditions (soil moisture, temperature, nutrients, pollutants, etc.) may adversely affect mycorrhizal symbiosis (14,33). Since mycorrhizal fungi exhibit ecological selectivity, cultural practices may need to be slightly adjusted to provide an environment conducive to fungus-host symbiosis. These observations also suggest

the importance of ectomycorrhizal fungi isolate selection for
use in nursery or greenhouse systems. The ultimate measure
of mycorrhizal inoculum effectiveness, however, is the seedling
growth response following outplanting.

14.4.2 Field performance of ectomycorrhizal planting stock

   The growth of the container-grown seedlings inoculated
with P. tinctorius was superior to that of noninoculated
containerized and bareroot seedlings two years after outplanting. After one growing season, leaf area and root system
length of the inoculated container-grown seedlings were 55
and 32% greater, respectively, than observed for the noninoculated containerized stock. Consequently, shoot growth of P.
tinctorius inoculated seedlings was superior to that of other
stock types. Farmer (11,12) emphasized the importance of
leaf area (photosynthetic potential) to vigorous root and
shoot growth following outplanting. Vigorous new root initiation following outplanting is essential if the oak seedling
root system is to meet the water and nutrient needs of the
shoot (19).

   Inoculation of bareroot seedlings with P. tinctorius did
not consistently increase root or shoot growth in the field.
Moreover, P. tinctorius ectomycorrhizae dramatically declined
on the bareroot stock during the first growing season. Few
new lateral roots were observed on the bareroot seedlings
during the study, and many older laterals were sloughing off.
Carpenter and Guard (3) concluded that transplanted bareroot
oaks have little success because they typically have few
laterals and the laterals present are "highly suberized" and
largely nonfunctional absorbing organs. Delays in the onset
of new root growth may have predisposed the bareroot seedlings
to water and nutrient stress which resulted in poor shoot
growth, extensive shoot dieback, and mortality.

   Little published information exists on the field performance of oak inoculated with specific ectomycorrhizal fungi.
Vlasov (43), Runnov (36), and Shemakhanova (40) reported
significant increases in shoot length, root collar diameter,

and shoot and leaf dry weight of outplanted mycorrhizal oaks. Shemakhanova (40) also observed that the effective root surface area of mycorrhizal oaks was twice that of noninoculated seedlings and reported 70% increases in seedling leaf area associated with abundant mycorrhizal colonization. The superior root development of ectomycorrhizal oaks is especially important in droughty, infertile soils, such as those common to upland oak sites in the Missouri Ozark region, since it enables the seedling to exploit a larger volume of soil for moisture and nutrients. Not only does the ectomycorrhizal seedling benefit from the enlarged root absorption area, but the fungal hyphae emanating from the root system further expands the pool of nutrients and water available to the plant.

Previous investigations (11,12) have suggested that relatively large bareroot oak seedlings (0.5 m in height, 50 g oven dry weight) are required to realize desirable growth rates following outplanting. The poor growth of the 1-0 bareroot seedlings following outplanting in this study (mean dry weights prior to planting of 17.0 and 13.8 g for the inoculated and noninoculated seedlings, respectively) supports this conclusion (9). However, the superior growth of even smaller container-grown seedlings (5.9 and 4.9 g for inoculated and noninoculated seedlings prior to planting, respectively) suggests that other factors such as root:shoot balance, root system length, number of primary lateral roots, and a relatively large leaf area (photosynthetic potential) may also be important in determining seedling growth potential and subsequent outplanting success. Johnson (19) demonstrated the superior capacity of container-grown seedlings, with relatively large numbers of primary lateral roots, to initiate rapid shoot and root growth. The development of a large leaf area is also necessary to provide the necessary carbohydrates to initiate and sustain an efficient root and shoot growth feedback system. The complementary relationship between root initiation and leaf area development of the container-grown

seedlings inoculated with P. tinctorius suggests continued
superior growth in comparison with the other stock types
outplanted in this study.

14.4.3 Ecophysiological characteristics of ectomycorrhizal planting stock

Both container culture and inoculation with P. tinctorius
significantly influenced seedling water relations following
outplanting. Throughout the first year, the container-grown
seedlings inoculated with P. tinctorius exhibited higher Ψ_{pd}
values than the other stock types. It is possible that this
pattern of Ψ_{pd} was attributable to the capacity of container-
grown and inoculated seedlings to utilize water not available
to bareroot or noninoculated seedlings (32,38), to equilibrate
more rapidly at night with available soil water (8), or both.

Midday water deficits were significantly lower if seedlings
were previously grown in containers or inoculated with P.
tinctorius. Owston (30) observed that bareroot seedlings may
be more likely to develop greater internal water deficits
because of their truncated taproots and relatively restricted
root system length; this was apparently the case in this
study. This improved capacity of the ectomycorrhizal seedlings
to moderate water stress is consistent with the pattern of
water stress observed in well-watered ectomycorrhizal and
nonmycorrhizal white oak seedlings observed in another study
(8). Beneficial effects on black oak seedling growth would
appear likely in this study, as Ψ_{sn} values of the noninoculated bareroot plants were low enough during the mild drought
to have impaired photosynthesis (17).

Analysis of soil-plant liquid flow resistance (R_{s-p})
revealed that containerization and ectomycorrhizal inoculation
facilitated liquid water movement from the soil to the leaves.
It was not possible to define precisely the portion(s) of
the soil-plant system (soil, soil-root interface, root interior,
root to leaf pathway) that possessed lower resistance(s).
However, soil moisture was relatively abundant throughout
the study period and therefore soil hydraulic conductivity

could be expected to be high (41). The container-grown
seedlings, with lengthy root systems, have been shown to have
large numbers of unsuberized roots (19,20). Hence the low
R_{s-p} value of the containerized seedlings may have been
partially attributable to large root absorptive areas. In
mycorrhizal roots it has been suggested that an increased
efficiency in water uptake (i.e., lower plant resistance) may
result from (i) increased absorbing surface resulting from
the presence of and penetration into the soil by hyphae,
(ii) reduced development of vapor gaps between soil particles
and root surfaces associated with reduced shrinkage in
mycorrhizal roots (32), (iii) functioning of the fungus as
a low-resistance pathway for water movement through the root
cortex, and (iv) stimulated root growth (8,38). It is
possible that stimulation of root growth as a result of
ectomycorrhizal infection could account for the large difference in R_{s-p} observed between inoculated and noninoculated
bareroot seedlings (8). It is more likely that other
mechanisms noted above may function to lower resistance from
the root-soil interface inward to the root interior. Fungal
hyphae have been shown to translocate water from "wet areas"
in the soil through areas of high soil moisture stress to the
seedling host (32,40). The significance of this phenomenon
was observed by Runnov (36) when nonmycorrhizal seedlings
ceased to grow under low soil moisture conditions, whereas
mycorrhizal plants continued to grow and maintain their
foliage during periods of severe water stress.

14.5 SUMMARY

The results of this study indicate that use of container-
culture and ectomycorrhizal inoculation significantly improve
(i) seedling root and shoot growth, (ii) water balance, and
(iii) soil-plant liquid flow resistance of black oak seedlings
outplanted in the Missouri Ozark region. Seedlings planted
without the benefit of an ectomycorrhizal association may
lack the growth potential necessary to compete with adjacent
vegetation.

316

The potential for improving the growth of outplanted oak seedlings through manipulation of mycorrhizae technology is promising. However, a greater understanding of ectomycorrhizal relationships is needed before further biological and economic benefits can be fully realized. Basic research on the physiological relationships of mycorrhizae should continue. Screening of candidate ectomycorrhizal fungi in nursery, glasshouse, and field trials is also greatly needed.

14.6 ACKNOWLEDGMENTS

The assistance of Ivan Sander, Paul Johnson, Galen Wright, Charles Dale, Ken Davidson, and Connie Campagna is gratefully acknowledged. This research was supported by the McIntire-Stennis Program and the USDA Forest Service North Central Forest Experiment Station under Cooperative Agreement 13-524.

REFERENCES

1. Bjeckford, P. R., R. E. Adams, and D. W. Smith. 1980. Effect of nitrogen fertilization on growth and ectomycorrhizal formation of red oak. For. Sci. 26:529-536.
2. Bowersox, T. W. and W. W. Ward. 1972. Prediction of advance regeneration in mixed-oak stands of Pennsylvania. For. Sci. 18:278-282.
3. Carpenter, I. W. and A. T. Guard. 1954. Anatomy and morphology of the seedling roots of four species of the genus Quercus. J. For. 52:269-74.
4. Dixon, R. K., G. T. Behrns, G. S. Cox, H. E. Garrett, J. E. Roberts, P. S. Johnson, and I. L. Sander. 1981. The influence of soil temperature on growth and ectomycorrhizal relationships of Quercus velutina seedlings. In Proceedings of the 3rd Central Hardwood Conference (H. E. Garrett and G. S. Cox, eds.), University of Missouri. p. 289-297.
5. Dixon, R. K., H. E. Garrett, J. M. Bixby, G. S. Cox, and J. G. Tompson. 1981. Growth, ectomycorrhizal development and root soluble carbohydrates of black oak seedlings fertilized by two methods. For. Sci. 27:617-624.
6. Dixon, R. K., H. E. Garrett, G. S. Cox, P. S. Johnson, and I. L. Sander. 1981. Container- and nursery-grown black oak seedlings inoculated with Pisolithus tinctorius: growth and ectomycorrhizal development following outplanting on an Ozark clearcut. Can. J. For. Res. 11:492-496.

7. Dixon, R. K., S. G. Pallardy, H. E. Garrett, G. S. Cox, and I. L. Sander. 1983. Comparative water relations of container-grown and bareroot ectomycorrhizal and non-mycorrhizal Quercus velutina seedlings. Can. J. Bot. 61:1559-1565.

8. Dixon, R. K., G. M. Wright, G. T. Behrns, R. O. Teskey, and T. M. Hinckley. 1980. Root growth and water relations of ectomycorrhizal white oak. Can. J. For. Res. 10:545-548.

9. Dixon, R. K., G. M. Wright, H. E. Garrett, G. S. Cox, P. S. Johnson, and I. L. Sander. 1981. Container- and nursery-grown black oak seedlings inoculated with Pisolithus tinctorius: growth and ectomycorrhizal development during seedling production period. Can. J. For. Res. 11:487-491.

10. Doak, K. D. 1955. Mineral nutrition and mycorrhizal association of bur oak. Lloydia. 18:101-108.

11. Farmer, R. E. 1975. Dormancy and root regeneration of northern red oak. Can. J. For. Res. 5:176-185.

12. _____. 1975. Growth and assimilation rate of juvenile northern red oak; effects of light and temperature. For. Sci. 21:373-381.

13. Garrett, H. E., G. S. Cox, R. K. Dixon, and G. M. Wright. 1979. Mycorrhizae and the artificial regeneration potential of oak. In Regenerating oaks in upland hardwood forests, John S. Wright Forestry Conference Proceedings (H. A. Holt and B. C. Fischer, eds.), Purdue University. p. 82-90.

14. Harley, J. L. 1969. The biology of mycorrhizae. 2nd edition. Leonard Hill.

15. Harley, J. L. and H. E. Smith. 1983. Mycorrhizal symbiosis. Academic Press.

16. Hewitt, E. J. 1966. Sand and water culture methods used in the study of plant nutrition. Commonw. Bur. Hortic. Plant Crops Tech. Commun. 22:547.

17. Hinckley, T. M., R. G. Aslin, R. R. Aubuchon, C. L. Metcalf, and J. E. Roberts. 1978. Leaf conductance and photosynthesis in four species of the oak-hickory forest type. For. Sci. 24:73-84.

18. Johnson, P. S. 1974. Containerization of oak seedlings for the oak-hickory region--a progress report. In Proceedings North American Containerized Forest Tree Seedlings Symposium (R. W. Tinus, W. I. Stein, and W. E. Balmer, eds.), Great Plains Agricultural Council Publication No. 68. p. 104-111.

19. _____. 1979. Growth potential and field performance of planted oaks. In Regenerating oaks in upland hardwood forests, John S. Wright Forestry Conference Proceedings (H. A. Holt and B. C. Fischer, eds.), Purdue University. p. 113-119.

20. _____. 1980. Oak planting in mid-south upland hardwoods: problems and prospects. In Proceedings Mid-South Upland Hardwood Symposium for the Practicing Forester and Land Manager (F. Shropshire and D. Sims, eds.), Harrison, Arkansas. p. 74-87.

21. Johnson, P. S. and H. E. Garrett (eds.). 1981. Workshop on seedling physiology and growth problems in oak planting. USDA For. Serv. Tech. Rep. NC-62. 26 p.

22. Marx, D. H. 1973. Mycorrhizae and feeder root diseases. In Ectomycorrhizae: Their Ecology and Physiology, (G. C. Marks and T. T. Kozlowski, eds.). Academic Press. p. 351-8?

23. Marx, D. H. 1979. Synthesis of Pisolithus ectomycorrhizae on white oak seedlings in fumigated nursery soil. USDA For. Serv. Res. Note SE-280. 6 p.

24. _____. 1979. Synthesis of ectomycorrhizae by different fungi on northern red oak seedlings. USDA For. Serv. Res. Note SE-282. 8 p.

25. Marx, D. H. and W. C. Bryan. 1975. Growth and ectomy- corrhizal development of loblolly pine seedlings in fumigated soil infested with the fungal symbiont Pisolithus tinctorius. For. Sci. 21:245-254.

26. Marx, D. H., W. C. Bryan, and C. E. Cordell. 1976. Growth and ectomycorrhizal development of pine seedlings in nursery soils infested with Pisolithus tinctorius. For. Sci. 22:91-100.

27. _____. 1977. Survival and growth of pine seedlings with Pisolithus ectomycorrhizae after two years on reforestation sites in North Carolina and Florida. For. Sci. 23:363-373.

28. Mexal, J. G. 1980. Aspects of mycorrhizal inoculation in relation to reforestation. N. Z. J. For. Sci. 10:208-217.

29. Olson, D. F. and R. M. Hooper. 1968. Early survival and growth of planted northern red oak in the southern Appalachians. USDA For. Serv. Res. Note SE-89. 3 p.

30. Owston, P. W. 1972. Cultural techniques for growing containerized seedlings. In Western Forestry Nursery Council and Intermountain Forest Nurseryman's Association Joint Meeting Proceedings. p. 32-41.

31. Quigley, K. L. 1971. The supply and demand situation for oak timber. In Oak Symposium Proceedings (D. E. White and B. A. Roach, eds.). USDA For. Serv. p. 30-36.

32. Reid, C. P. P. 1979. Mycorrhizae and water stress. In Proceedings of the Symposium: Root Physiology and Symbiosis (A. Riedacker and J. Gagnaire-Michard, eds.). CNFR. p. 392-408.

33. Riffle, J. W. and D. M. Maronek. 1982. Ectomycorrhizal inoculation procedures for greenhouse and nursery studies. In Methods and Principles of Mycorrhizal Research (N. C. Schenk, ed.). Am. Phytopathol. Soc. p. 147-156.

34. Rowse, H. R. and D. A. Phillips. 1974. An instrument for estimating the total length of a root in a sample. J. Appl. Ecol. 11:309-314.

35. Ruehle, J. L. 1980. Ectomycorrhizal colonization of container-grown northern red oak as affected by fertility. USDA For. Serv. Res. Note SE-297. 5 p.

36. Runnov, E. V. 1955. Experimental introduction of mycorrhizae into oak sowings in arid steppe. In Mycotrophy in Plants (A. A. Imshenetskii, ed.). Israel Program of Scientific Translations. p. 174-186.

37. Russell, T. E. 1971. Seeding and planting upland oaks. In Oak Symposium Proceedings (D. E. White and B. A. Roach, eds.). USDA For. Serv. p. 49-54.

38. Safir, G. R., J. S. Boyer, and J. W. Gerdemann. 1972. Nutrient status and mycorrhizal enhancement of water transport in soybean. Plant Physiol. 49:700-703.

39. Sander, I. L., P. S. Johnson, and R. F. Watt. 1976. A guide for evaluating the adequacy of oak advance reproduction. USDA For. Serv. Gen. Tech. NC-23. 6 p.

40. Shemakhanova, N. M. 1962. Mycotrophy of woody plants. U.S. Dept. Commerc. Transl. TT66-51073.

41. Slayter, R. O. 1967. Plant-water relationships. Academic Press.

42. Tinus, R. W. and S. McDonald. 1979. How to grow tree seedlings in containers in a greenhouse. USDA For. Serv. Gen. Tech. Rep. RM-60. 256 p.

43. Vlasov, A. A. 1955. Importance of mycorrhiza for forest trees and procedures for its stimulation. In Mycotrophy in Plants (A. A. Imshenetskii, ed.). Israel Program of Scientific Translations. p. 107-117.